Gender, Power and Politics in Agriculture

Jemimah Njuki • Hale Ann Tufan
Vivian Polar • Hugo Campos
Monifa Morgan-Bell
Editors

Gender, Power and Politics in Agriculture

Revisiting Theory and Practice

 Springer

Editors
Jemimah Njuki
Chief of Economic Empowerment
UN Women
New York, NY, USA

Vivian Polar
Leader of the Social and Nutritional
Sciences Division
International Potato Center
Lima, Peru

Monifa Morgan-Bell
School of Integrative Plant Science
Cornell University
Ithaca, NY, USA

Hale Ann Tufan
Associate Professor at School of Integrative
Plant Science
Cornell University
Ithaca, NY, USA

Hugo Campos
Deputy Director General for Science
and Innovation
International Potato Center
Lima, Peru

ISBN 978-3-031-60985-5   ISBN 978-3-031-60986-2   (eBook)
https://doi.org/10.1007/978-3-031-60986-2

This Springer imprint is published by the registered company Springer Nature Switzerland AG
The registered company address is: Gewerbestrasse 11, 6330 Cham, Switzerland

If disposing of this product, please recycle the paper.

We dedicate this book to all of those who strive for gender equity. May this book be a source of inspiration and knowledge in the path toward a more equitable world.

Foreword

The growing pressure and challenges faced by the agricultural sector are as large and complex as the social, organizational, and political challenges the world faces. In an evolving agricultural landscape, it is crucial to explore critical social and gender issues surrounding power and politics in the sector. The book "Gender, Power and Politics in the Agricultural Sector" offers a compelling journey through 11 thought-provoking chapters, which invite readers to revive feminist debates, challenge prevailing growth models, and explore the perpetual struggles faced by gender researchers in agriculture.

Throughout the book, the authors question assumptions that underpin agricultural research and development, making a case for more feminist and rights-based development models. This analysis is a wake-up call, urging us to question the existing paradigms. The authors challenge us to move beyond mere tokenism, pushing for a profound reevaluation of the role of women in agriculture beyond the instrumentalist lens.

In analyzing the contextual factors that limit the connection between agricultural research and feminist methods and theory, the authors highlight the marginalized role of gender researchers that remain on the fringes of academia and research. The book elaborates on how gender and women's studies as an interdisciplinary academic field is faced with challenges of recognition, legitimacy, funding, and visibility. Across several chapters, the book sheds light on the importance and potential benefits of a more inclusive and supportive environment for those striving to bridge the gender gap in agricultural studies. Furthermore, the instrumentalization of women and feminist agendas is analyzed through the proliferation of tools for gender analysis and mainstreaming in agricultural research processes. This analysis evidences the dangers of relying solely on technological solutions without addressing the underlying power dynamics. It encourages readers to question the notion of progress that neglects the nuances of gender and power in agriculture.

A fourth element of analysis across the book is the nature of organizations and the reproduction of the status quo. Examples of "Self-replicating institutions" and mechanisms through which patriarchy is baked into organizational culture in agriculture are presented. The analysis delves into the roots of institutional and organizational structures that perpetuate gender inequalities. By exposing these systemic issues, the authors challenge readers to recognize and dismantle pervasive patriarchal norms and persistent institutional sexism, calling for collective action to break these patterns.

As readers approach the final section, they will find a hopeful exploration of possibilities for change. The authors present a compelling case for why this book is not just a critique but a catalyst for transformation. By identifying emerging opportunities, the book serves as a guide for trajectories. "Gender, Power and Politics in the Agricultural Sector" is not just a book; it is a call to action that encourages us to rethink our assumptions, question prevailing norms, and actively engage in reshaping the future of agriculture and the institutions that lead research and development processes in the sector.

Ismahane Elouafi
Executive Managing Director
CGIAR

The original version of the book has been revised. A correction to this book can be found at https://doi.org/10.1007/978-3-031-60986-2_12

Acknowledgments

First and foremost, we would like to thank all the authors and the organizations they are associated with for their willingness to contribute their time and invaluable experiences. This would not have been possible without their commitment and dedication to advance gender equality in the agricultural sector.

Drs. Jacqueline A. Ashby, Helen Hambly, and Graham Thiele provided critical review of the chapters, and Jeffrey Bentley edited and provided meticulous support in referencing and formatting. Without their support this process would not have been possible.

We would like to wholeheartedly thank the Bill & Melinda Gates Foundation for enabling us to publish our book as Open Access, as this would allow it to reach a wide audience.

We are very grateful to our editor Joao Pildervasser and the Springer team, for all the patience, support, and work behind the scenes. We highly valued their outstanding support to advance the book toward publication.

Personal Acknowledgements

I (**Jemimah Njuki**) thank the two organizations that I have worked for during the writing of this book, the International Food Policy Research Institute and UN Women, for their support and for allowing me the time to pursue this important work with my co-editors. To my co-editors Hale, Vivian, Hugo, and Monifa, it has been such a pleasure working with you on this book. Thank you for your patience through the transition between my two jobs. And to my two daughters, Tamara and Rachel, who inspire me every day to work toward ensuring that every woman, and every girl lives a dignified, fulfilling, and rewarding life.

I (**Hale Ann Tufan**) strive every day to make a better world for my kids, so my gratitude goes to them for inspiring me to do so. A big thank you to Burak for supporting me through a busy career by taking on a myriad of care duties. My parents and sister are my lifeline, so thank you for being there. The laughter, learning, and letting go that was central to our editorial process with Jemimah, Vivian, Hugo, and Monifa has set the bar high for any book endeavor, thank you all for those moments.

I (**Vivian Polar**) am grateful to my mother Justina, who inspired my professional growth and continues to inspire my life every day. I am grateful to my husband Luis and to my children who are continuous source of inspiration and support. A heart felt thank you to my co-editors for being a source of motivation and encouragement. I dedicate this book to the women farmers of the world and to those who tirelessly strive for gender equity.

I (**Hugo Campos**) am very grateful to the International Potato Center for allowing me to pursue my professional dreams and interests, and to Jemimah, Hale, Vivian, and Monifa, as editing this book has been such a joyful, fulfilling experience, and grateful for their forgiveness about my shortcomings. I am indebted beyond words to my wife Orietta for tolerating yet another writing assignment. This book is dedicated to her since without her tireless love, encouragement, and support, neither this book nor any other of my modest professional accomplishments would have ever occurred. The expert editorial advice of Miss Noelia Campos is much appreciated.

I (**Monifa Morgan-Bell**) express my gratitude to Cornell University that continues to uphold its founder's motto of "an institute where any person can find instruction in any study." Cornell has opened so many doors and opportunities to me, for which I am truly grateful. I would like to thank Hale Tufan, who continues to be a great mentor, leader, and friend, for including me in this work. It has been life changing. To my co-editors, Jemimah, Vivian, and Hugo, thank you for sharing your wealth of knowledge, insights, and laughter on this amazing journey.

Closing Remarks

To the readers of our book, all its coauthors share the hope that the insights shared will inspire you to change the status quo and generate new alternative pathways toward gender equity and well-being. That will mean this book has been a success.

New York and Lima, September 2024.

Jemimah Njuki, Hale Ann Tufan, Vivian Polar, Hugo Campos, and Monifa Morgan-Bell

Contents

About the Editors

Jemimah Njuki is a Kenyan national with over 25 years of experience working on gender equality in agriculture and on women's economic empowerment. She currently serves as the Chief, Economic Empowerment at UN Women in New York. She has previously worked at the International Food Policy Research Institute, the Canadian International Development Research Centre, the International Livestock Research Institute, the International Centre for Tropical Agriculture, and CARE USA. She holds a PhD in Development Studies from Sokoine University, and a bachelor's degree in Dairy, Food Science and Technology.

Hale Ann Tufan is a Turkish-American dual national currently an Associate Professor at the School of Integrative Plant Science, Cornell University. Her work focuses on critical analysis of crop improvement research, conceptualizing inclusive crop improvement approaches and training, and gender research around crop breeding. She was awarded the Borlaug Field Award in 2019 to recognize her work to ensure women farmers and researchers are fairly represented in agricultural research for development. She has a PhD from the John Innes Centre (UK), experience in international agricultural research for development program management, and gender capacity development.

Vivian Polar is a Bolivian national, with over 20 years of professional experience working with women and indigenous peoples. She currently serves as Leader of the Social and Nutritional Sciences Division at the International Potato Center in Lima, Peru. She holds a PhD in Social Sciences from the School of Oriental and African Studies, University of London; an MSc in Sustainable Rural Development from Tomas Frias University in Bolivia; and a degree in Agriculture from Universidad Mayor de San Andrés in Bolivia.

Hugo Campos is a Chilean national with over 30 years of professional international experience in several continents in the private sector and the Consultative Group on International Agricultural Research (CGIAR), serves as Deputy Director General for Science & Innovation at the International Potato Center in Lima, Peru.

He holds a PhD from the John Innes Centre (UK), a MBA from Universidad del Desarrollo (Chile), and a Professional Certificate in Innovation & Entrepreneurship from Stanford University (USA), and has ample experience in innovation, research management, and plant breeding. Keen writer, the books he has published/edited have been sold/downloaded over 600,000 times to date.

Monifa Morgan-Bell is a Jamaican-American dual national who currently works as a Research and Administrative Assistant at Cornell University. She previously worked for the Innovation Lab for Crop Improvement (ILCI) in the Department of Global Development in the College of Agriculture and Life Sciences at Cornell University, as the program assistant. She was the recipient of the 2023 Award for Staff Integrity, for her work with the Thomas Wyatt Turner Fellowship Program and their partnership with Minority Serving Intuitions (MSI) at Cornell University.

Chapter 1
Introduction: The Politics of Gender and Agriculture

Jemimah Njuki (iD)**, Hale Ann Tufan** (iD)**, Vivian Polar** (iD)**, Hugo Campos** (iD)**,
Monifa Morgan-Bell, and Vicki Wilde**

Abstract As researchers and practitioners at various stages of our careers and from diverse disciplines, with many decades of collective experience, we have witnessed an evolution in the theory and practice of gender and agriculture. What compelled us to put this book together was a growing sense of frustration from the global community of gender and agriculture researchers with the pervasive co-option of the "gender agenda", along with a de-politization of its critical theories and interventions with roots in radical change. We recognize this book is a synopsis of only some possible perspectives, but in reaching out to authors to contribute, it was our aim to create an opportunity to publish the things they felt are urgent today, but perhaps felt were too disruptive, challenging or without enough space in the mainstream body of literature. In what follows, we question some of the assumptions that underpin agricultural research and development, make clear our support for the nascent rise of more feminist and rights-based development models, and set the scene for this book. We call for a reset.

As researchers and practitioners at various stages of our careers and from diverse disciplines, with many decades of collective experience, we have witnessed an evolution in the theory and practice of gender and agriculture. What compelled us to put this book together was a growing sense of frustration from the global community of

J. Njuki (✉)
United Nations Entity for Gender Equality and the Empowerment of Women and Girls, UN Women, New York, NY, USA
e-mail: jemimah.njuki@unwomen.org

H. A. Tufan
Global Development, Cornell University, Ithaca, NY, USA

V. Polar · H. Campos
International Potato Center, Lima, Peru

M. Morgan-Bell
School of Integrative Plant Science, Cornell University, Ithaca, NY, USA

V. Wilde
Bill and Melinda Gates Foundation, Seatle, Washington, USA

© The Author(s) 2025
J. Njuki et al. (eds.), *Gender, Power and Politics in Agriculture*,
https://doi.org/10.1007/978-3-031-60986-2_1

gender and agriculture researchers with the pervasive co-option of the "gender agenda", along with a de-politization of its critical theories and interventions with roots in radical change. We recognize this book is a synopsis of only some possible perspectives, but in reaching out to authors to contribute, it was our aim to create an opportunity to publish the things they felt are urgent today, but perhaps felt were too disruptive, challenging or without enough space in the mainstream body of literature. In what follows, we question some of the assumptions that underpin agricultural research and development, make clear our support for the nascent rise of more feminist and rights-based development models, and set the scene for this book. We call for a reset.

1.1 Re-kindling Feminist Debates on Gender Equality in Agriculture: Can We Move Beyond Instrumentalist Growth Models Built on Women's Labor?

The roots of gender and agriculture scholarship originated in feminist movements involving rural women. In turn, feminist movements generated feminist theory. Broadly, feminist theory calls for the replacement of the patriarchal order with a system that emphasizes equal rights, justice, and fairness. What this means and how it should be achieved is argued through many different feminist theories. Liberal feminists cite women's oppression as rooted in social, political, and legal constraints (Tong 2009; Giddens 2001; Maynard 1995), while radical libertarian feminists hold that the patriarchal system that oppresses women must be eliminated and that women should be free to exercise total sexual and reproductive freedom (Thompson 2001; Echols 1989). Global and transnational feminists stress the universal interests of women worldwide, through solidarity (Ferree and Tripp 2006). Ecofeminists focus on the connection among humans to the natural world. Intersectional feminism draw attention to how social identities form an intersecting experience of oppression (Gaard 1993, 2017; Shiva and Mies 2014; Plumwood 1986). De-colonial and third world feminists challenge power structures that originated in colonialism (Ossome 2022; Mehta 2020). Taken together, feminist theory has examined virtually all structures, systems, and disciplines, and has challenged traditional ontological and epistemological assumptions about the nature of research. Nevertheless, feminist theory has scarcely permeated the world of agriculture research. The application of feminist theory into research and practice, with positive impacts on the livelihoods of women and girls involved in smallholder agriculture in the global South, has not been adequate and gender and agriculture programs have fallen short in several aspects. Farhall and Rickards (2021) in their paper on "The Gender Agenda in Agriculture for Development and Its (Lack of) Alignment With Feminist Scholarship" propose that there are problematic narratives that instrumentalize women in the name of sustainable agricultural development. This instrumentalization of women in agriculture and agricultural research for development, recognizing women's labor solely as a means to increasing productivity for example, reflects antiquated understandings arising from inadequate integration of feminist theory into research practice (Batriwala and Dhanraj Undated; Cornwall 2018).

There has also been a growing donor momentum leading to policies to "invest" in women in agriculture, with good reason. The recently released FAO report on the status of women in agrifood systems (FAO 2023) states that more than one out of three women globally work in the agri-food sector, with the proportion in sub-Saharan Africa and Southeast Asia being much higher than the global average, where more than two out of three women are involved in the agri-food sector. Framing women as important for the agri-food system because of their labor contribution however only narrowly ties rural women's roles to economic growth and productivity. Since the launch of the seminal World Bank report on Gender and Development (WB 2012), the focus on women as drivers of economic growth has risen. The argument, for example as articulated by the Food and Agriculture Organization (FAO 2011) is that if we invest in women in agriculture by providing them with more inputs such as fertilizer and pesticides or more credit, the world's productivity will increase by 20–30 percent.

What we have learned over the past decade is that "cutting and pasting" women and girls into productivity growth models for agriculture that ignore the systemic inequalities of a global economy that continues to perpetuate patriarchal and neo-colonial markets, gender blind financial policies, and discriminatory social norms, does not result in inclusive growth or reduce inequality. Widespread reliance on productivity growth models may in fact worsen the odds for poor women, their families, and their communities, despite good intentions. Simply targeting women and girls to produce more food or to provide them with more safety nets does not address the structural problems that are also affecting their rights and aspirations or their access and control to key resources. Paradoxically, it can actually perpetuate the status quo. Much of current development practice includes a focus on women to undergird intensive productivity goals rather than promote a more feminist and rights-based development agenda to tackle the root causes of poverty and inequality.

Fortunately, there are encouraging pockets of progress. Of note is the rise of gender transformative approaches that are challenging unequal power relations and supporting women as agents of their own development priorities. There are promising examples of applying feminist theories to challenge patriarchy and develop inclusive and resilient models for agricultural growth and sustainable food systems. The FAO 2023 data makes clear that if we were to focus more of our policies, strategies, programs, and interventions on gender transformative outcomes, the entire agrifood system would benefit. Where women have the agency to make their own decisions and pursue new life options, without fear of violence, there are increases in productivity, food security, nutrition, and most especially, resilience to climate change.

1.2 From Margin to Marginalized: The Perpetual Plight of Gender Research and Researchers in Agriculture

A second reason for inadequate presence of feminist theory in agricultural research practice is that gender research and researchers remain on the fringes of academia and research. Gender and Women's Studies as an interdisciplinary academic field itself is faced with challenges to its recognition, legitimacy, funding and visibility. It is constrained by a broader context of socio-political, religious, and patriarchal resistance to the kinds of knowledge and interventions produced by feminist discourse and gender critiques.

In agricultural research institutions, traditionally, a similar marginalization is felt keenly by gender researchers. Low resource allocation and inadequate staffing of gender research programs was justified by the prominence of biophysical or economic research agendas. Power dynamics in agricultural research teams and institutions tended to relegate gender research and researchers to the margins (Cullen et al. 2023). At the same time, social science, gender research included, often was viewed as service provision to understanding other aspects of research and agricultural growth (Cernea 2005). Locked into marginal service roles with little power, a large proportion of the early gender (and some current) research uses gender (conflated with sex) as an analytical "co-variate" for example to understand low adoption of improved technologies. This type of research suggests a failure to consider how the study of gender can provide highly relevant analyses of the social processes that underpin development, and how it sheds light on the social norms that underlay the power, knowledge, institutional, and market dynamics that will determine whether interventions will be inclusive or sustainable.

Feminist theory also challenges the social sciences through critiques of "existing fantasies of objective knowledge produced by autonomous subjects" (Hemmings 2012). Critical social theories and frameworks emphasize how ontological and epistemological orientations inform theory, as well as methodology, methods, and analysis. But there has been only sparse application of critical social theory and feminist theory in agricultural research. Examples of critical social theories, frameworks, and methodologies that are attentive to oppressive social structures and empowerment for minoritized groups include community cultural wealth (Yosso 2005), critical race theory in education (Gillborn and Ladson-Billings 2009; Ladson-Billings and Tate 1995; Lynn and Dixson 2013; Patton 2016; Solorzano 1997), decolonizing and anticolonial methodologies (Bhattacharya 2019; Patel 2014; Smith 2012), funds of knowledge (Kiyama and Rios-Aguilar 2017), funds of identity (Esteban-Guitart 2016), racial capitalism (Robinson 2020), and racial formation theory (Omi and Winant 2014).

The recent uptick in gender transformative approaches in agriculture has been a welcome development, and so has been the increasing focus on intersectionality. Evans-Winters and Esposito (2019) explain how feminist and critical race frameworks such as intersectionality (Crenshaw 1989) are useful in critiquing social order and hierarchy. Researchers, including those in the agriculture sector have started to draw upon intersectionality as an analytical tool in the examination of (and resistance to) systems of power, privilege, and domination. Here too there may be a risk of co-option as the concept of intersectionality becomes another "double-covariate" in simplistic analysis, divorced from critical analysis of systems of oppression. Fortunately, we see a growing number of organizations use feminist research as a mode of analysis to identify and challenge the ways that hegemonic, patriarchal systems of domination impact women and communities broadly, often centered on directly supporting rural women's local organizations (Njuki et al. 2022).

1.3 Letting the Tools Do the Hard Work and Training for Instrumentalism: Mainstreaming Mediocracy in Agriculture

The third reason, related to the instrumentalization of women and feminist agendas, is the increased focus on tools and training for gender analysis and gender mainstreaming. Given the paucity of gender researchers in most agriculture research institutions, these were intended to increase awareness among non-experts, with limited success. Tools often came at the expense of a deeper interrogation of social and economic power dynamics. Since the release of the seminal book on Gender Planning and Development: Theory, Practice and Training by Caroline Moser in 1993 the growth of gender and development frameworks and associated tools proliferated. According to the author '...many of those committed to integrating gender into their work at policy, programme, and project levels still lack the necessary planning principles and methodological tools... planners require simplified tools which allow them to feed the complexities of specific contexts into the planning process'. (Moser 1993, pp. 5). These tools and frameworks were therefore intended to meet the needs of development practitioners, trainers, researchers, and students for conducting gender analysis and for integrating gender in development programs. And while these tools and frameworks have been useful, they have not sufficiently addressed the power relations and underlying structural causes of gender inequalities. Their application has also meant that even those that do not understand gender, can cherry-pick which tools they use to meet certain purposes without a critical understanding of the context in which these tools are used. And as Naila Kabeer says: 'No set of methods are in themselves sensitive to differences and inequalities between men and women; each method is only as good as its practitioner' (Kabeer 1995, pp. 112). There also has been widespread tendency of development practitioners to by-pass theory—in this case feminist theories and theories around gender relations—in the mistaken assumption that the tools are sufficient, forgetting that the tools are purely technocratic deprived of their political dimension. The same can be said for gender training. Critical scholars have long berated short checklist training events (Mukhopadhyay 2014) or safe training events that do not alienate or challenge participants (Chant 2012). Gender training with a focus on skills building, without challenging participants leads to oversimplification of feminist theoretical conceptualizations of power (Mukhopadhyay and Wong 2007). Ferguson (2015) dissects gender training events, berating their complete depolitization, for fear of being perceived as political. Moving away from gender awareness to examining root causes of gender inequality, necessarily requires that we expose researchers and practitioners to understand their roles in reinforcing them.

1.4 Self-Replicating Institutions: Patriarchy Baked into Organizational Culture in Agriculture

The fourth reason is the nature of organizations to reproduce the status quo. While there is an understanding of patriarchy as a way that society is organized, this has rarely been applied to organizations. Carrying out transformative gender research and analyzing power dynamics and unequal power relations and doing that in a patriarchal organization is problematic. These organizations tend to recreate and reproduce systems that exclude women, and gender diverse individuals who sit outside of these masculine institutional norms. Men's economic domination of agriculture and the perception of agriculture as a masculine space has been studied in the Global North (Saugeres 2002), possibly permeating the Global South through colonial expansion, despite the important role that women play in agriculture globally. Even today, the common perception of and discourse surrounding farming remains centered on the rugged, nature-taming man (Shisler and Sbicca 2019). The patriarchal system of accumulation of wealth has historically limited women's access to farmland and when they do have access, its control is often influenced by other men (Alsgaard 2012; Carter 2017).

Current agricultural technology and research started in the context of patriarchal institutions created to foster agricultural development after the second World War and continue to operate under commodity maximization paradigms intended to drive down global hunger. The agricultural research institutions originally were staffed by and for men and driven by assumptions that reflect the values and life situations of men and of idealized masculinity (Merrill-Sands et al. 1999), that have replicated over time through perceived workplace norms of success, commitment and leadership that entrench gender segregation and inequity in the workplace. Global movements by women in science, and dedicated efforts from within the institutions of agriculture research have built new on-ramps for women researchers and leaders, and more resources for gender research, but many challenges remain.

1.4.1 Why This Book? Emerging Opportunities for a Reset

Recent developments are putting a spotlight on gender equality and the rights of women and girls as a matter of urgency. Three years after the start of the pandemic, women's labor force participation rates remain suppressed in comparison to men's and the gender gaps have widened. It is estimated that the COVID-19 pandemic pushed an additional 47 million girls and women into extreme poverty, reversing decades of progress. Data from 40 countries show that 36% of women stopped working during the pandemic compared with 28% of men. Getting enough to eat has become more difficult too. In 2021, at least 150 million more women than men were experiencing food insecurity (FAO 2023).

The pandemic exploited pre-existing gender inequalities and increased the prevalence of gender-based violence. Women faced higher levels of harassment and exploitation both within their households and in their work environments. Additionally, women often lacked access to safety nets, healthcare services, and social security systems and therefore were further marginalized.

Women, who make up a significant proportion of small-scale farmers and food producers, faced challenges in accessing inputs, such as seeds and fertilizers, as well as markets to sell their produce. Reduced market demand, lower prices, and limited access to markets resulted in financial losses for many women farmers. The loss of income undermined their economic empowerment and made it harder for them to invest in their farms, access financial services, and participate in decision-making processes. With lockdowns and restrictions, women in agriculture often took on additional responsibilities. They had to manage household chores, childcare, and other domestic tasks while still fulfilling their agricultural roles. This increased workload put a strain on their physical and mental well-being and limited their time and energy for other activities, such as education or pursuing alternative income sources.

And yet, most COVID responses failed to address the structural barriers that led to women being disproportionately and negatively impacted. Policy responses also failed to acknowledge the important role that women, as farmers, processors, traders, and agricultural workers played in sustaining food systems. The COVID-19 Global Gender Response Tracker developed by UN Women and UNDP monitored responses taken by governments worldwide to tackle the pandemic and highlighted those that integrated a gender lens. It captured two types of government responses: women's participation in COVID-19 task forces and national policy measures taken by governments. Out of 4968 measures analyzed, only 1605 (32%) were identified as gender sensitive, and out of a total of 3099 social protection and labor market measures adopted in response to the pandemic, only 12% targeted women's economic security and only 7 per cent provided support for rising unpaid care demands. The post recovery must address the structural barriers as well as the wellbeing of women in ways that challenge institutions to embed gender equality. And there are lessons learnt from the pandemic. For example, analysis shows that countries with robust public services and gender-responsive social protection systems were in a better position to respond, while others had to improvise, under pressure, and with varying degrees of success.

Today we see the risk of a similar dynamic playing out in climate crisis responses and climate negotiations. While women's presence in local climate-smart agricultural responses is associated with better resource governance, conservation outcomes, and disaster readiness, women made up less than 34% of country negotiating teams at COP27. And of the 110 world leaders who were present, only 7 were women. Promoting women's leadership, ensuring their equal participation in decision-making processes, and investing in women's education and economic empowerment are crucial steps towards building gender-equitable climate-resilient societies. Another crucial step is ensuring that agriculture research institutions make it a priority to design climate-smart services, information, products and technologies that well serve rural women's priorities.

This all culminates to a moment of reconsideration. Is it time for a reset? The Food Systems Gender Stock Take that took place in Rome in 2023 opened dialogue and possibility for more radical change among food and agriculture organizations worldwide. Even large multinational organizations like FAO, with a history of narrowly focusing on gender to increase productivity, today promotes intersectionality, gender transformative approaches and gender justice in their latest status of women in food systems report (FAO 2023). Are we seeing a resurgence of feminist theory peering through agricultural research and practice? It is at this juncture that this book becomes not only relevant but timely.

It is our aspiration that the book will not only encourage open reflection and debate on the topics it covers, but also encourage agricultural research and practice to more readily pursue feminist goals and fulsomely support the deep, challenging work of gender transformative climate resilient food systems.

1.4.2   What the Book Contributes

This book explores the intersection between feminist theory, applied methodology and practice on gender and agriculture. We seek to address, to varying degrees, the four reasons set out in the introduction section, as disjuncture causing disconnect between feminist theory and agricultural research for development practice. The book is informed by the fact that such areas oftentimes work isolated from one another and that theoretical advancements may not be informing practice. Also, advancements in practice are not contributing as expected to reframe our theoretical thinking and methodological advancements. As a result of these divisions, within the agriculture sector, the language of gender integration often ignores the politics and the power dynamics involved in gender studies in this and other sectors. The book explores such neglected topics as power dynamics, masculinities, gendered social norms, intersectionality, which although are more common in other sectors such as health, nutrition, human rights, are still not mainstream in how gender and agriculture is researched, practiced and taught despite them being fundamental to our understanding of the interplay between gender, the agriculture sector and rural livelihoods.

Chapter 2 of the book by Hale Tufan and co-authors focuses on unpacking dominant narratives around gender and agricultural development and how they relate to critical feminist thought, between 1970–2020. The authors start with a framing chapter to outline a heuristic approach- including rendition of a gender tree and summarize major strains in feminist theoretical frameworks to guide the reader. They then present two historical narratives: *The undying allure of liberal feminism: Gender and agricultural development,* and *Critical approaches to land and gender justice.* The authors relate these two narratives showing how the mainstream narratives in agricultural development largely continued to address the consequences of gender inequality, while more critical approaches focused around land rights and access, challenge and target the root causes of gender inequality. In conclusion, they bring attention to the need for putting these two narratives more regularly in conversation, for mainstream gender and agriculture narratives to move closer to addressing root causes of inequality through gender transformative approaches, community focused interventions and a focus on gender norms and masculinities.

In Chapter 3 on power and politics in agriculture, Kansanga and Dinko use mechanization to show how agriculture has and still remains a masculine space despite the roles that women play in the sector. The authors argue that while mechanization has emerged as a key pathway for improving agriculture in the Global South, the materialization of its perceived benefits has been hindered by gender inequalities in technology use. This chapter shows that masculinization of agricultural technologies is not just a product of the widely discussed role of structural factors in local agrarian spaces, but an element of the very design/engineering of mechanized technologies that attribute masculine traits to them, and the consistent deployment of gender (in)sensitive agricultural programs by governments and development partners. The authors conclude that governments and development partners are viable intermediaries who are well positioned to channel feedback on the needs of women to the agricultural machinery industry upstream, while ensuring a gender-sensitive deployment of mechanization services downstream.

In Chapter 4, Jacqueline Ashby interrogates participatory research with a focus on *who* has agency and power in key decisions about whose knowledge is legitimate, what to research, how and at what cost. The chapter looks at what happened to gender in participatory research in science bureaucracies and the international agricultural research and development system initiated in the 1970s by the Ford and Rockefeller Foundations, to improve the productivity of staple crops in low-income countries. It examines how participatory research in agriculture and gender within it have been influenced by four features or "key drivers" structured by this institutional setting. First the institutional culture provided by science bureaucracies and the everyday practice of normal agricultural science that led to a piece-meal approach and dilution of leadership for participatory research and gender within it. Second, the incorporation of participatory research with the objective of improving supply-side, research efficiency. The third driver is called containing empowerment and reflects the preoccupation with inclusion rather than social justice. The fourth driver is the development of methods and tools which allowed participatory research pioneers in science bureaucracies to find legitimacy in the scientific model rather than in action research for transformative change. The chapter concludes that all four drivers have a common function: the appropriation of participatory and feminist ideology in ways that made it both palatable to powerful stakeholders, principally the leadership of public sector, agricultural research bureaucracies and their donors. This lead to the de-legitimization of transformative approaches of gender mainstreaming and participatory research as practiced in agricultural research.

Franz Wong, in Chapter 5 of the book speaks the unspoken to explore how masculinity became constitutive of AR4D through a historical gender analysis of Norman Borlaug and his influence on wheat breeding. Using the example of wheat agronomy in the early days of international AR4D, this chapter explores how AR4D practices are historically gendered and foundational to the maintenance of male-dominated agricultural research. In other words, AR4D not only reproduces gender bias but performs gender as a knowledge enterprise. The chapter further interrogates how agriculture research for development was constituted and traces this back to the history of Norman Borlaug, the 1970 Nobel Peace Prize recipient who, along with other scientists, were the lead researchers for agricultural innovations and what has become known as the "Green Revolution". Drawing on feminist technoscience studies and hegemonic masculinities, Wong offers us an interesting lens through which we can observe how agricultural research for development practice reproduces its own "gendered-ness".

In Chapter 6. Jemimah Njuki and co-authors focus on navigating the patriarchal politics of gender in agriculture research and practice focusing on how research organizations are themselves patriarchal in nature. The chapter argues that the issue of patriarchy in research has been approached as a way in which society is organized and as a topic of research and very rarely has the field of research applied this term to organizations and specifically the social and organizational context of how research is done and the power dynamics that govern the organizations carrying out research. The chapter goes on to define a patriarchal organization by identifying four characteristics (i) a fixed division of labor or segregation of tasks, (ii) hierarchy of offices, (iii) a set of rules governing performance, and (iv) a dominant figure or ruler with almost unlimited authority and power. It uses this characterization and applies them to show how research organizations exhibit some forms of these patriarchal characteristics. The chapter presents two case studies to that demonstrate possible avenues for change. The last section of the paper provides reflections on how research organizations can break these patriarchal tendencies and provides a vision of organizations where women are valued and where gender research is accorded the same value as other research.

In Chapter 7, Polar and Poole explore the instrumentalization of women's empowerment in agricultural research, with particular attention on critically examining how the concept of empowerment has become understood as an externalized process that can be bestowed on women through production-oriented interventions. The chapter explores the evolution of the concept of power and its multiple manifestations through concrete examples that analyze how the current use and assessment of empowerment has deviated from the original intent of empowerment around building agency and disrupting power dynamics. The chapter proposes a feminist and transformative conceptualization and operationalization of empowerment that explicitly analyzes manifestations of power-over to foster change in social and institutional structures that address research and development in the agricultural sector.

Cullen et al., in Chapter 8 unveil how the practice of agricultural research for development has been depoliticized through "the tyranny of tools" and the ways in which the application of gender analysis frameworks and tools has been done devoid of power analysis and a political economic analysis of the sector. The authors examine the emergence and development trajectory of analytical frameworks, and gender tools, that while intended to understand and address the challenges and inequities shaping women's participation in agriculture, often do little to address the structural inequality that women are embedded in. Tool-led gender analysis within agricultural projects, they argue, tends to detach tools from their theoretical frameworks, ignoring the structural and socio-political obstacles to gender equality in specific contexts, and, views tools as silver bullets for addressing 'gender problems' whilst primarily serving technical agendas. The co-option, sanitization and depoliticization of gender tools is partly the result of social scientists having to fit within institutional systems dominated by certain scientific logics, frameworks, disciplinary orientations, and social norms. In conclusion, the authors recommend that meaningful attempts to facilitate empowerment and transformation should be based on politically informed, contextualized understandings that are relevant to people's lived realities, rather than concepts, tools and data that are externally constructed and applied by outsiders to meet normative scientific, donor and development agendas.

In Chapter 9 Tavenner et al. focus on a topic that is, by and large, overlooked: intersectionality. The authors trace the origins of intersectionality in Black feminist thought, as a social theory to account for and better understand multiple and compounding identities and how they influence marginalization and privilege. Further, they outline the state-of-the-field on intersectional analyses in agriculture research for development and how they are situated within wider feminist mainstreaming in international development trends. Returning to the thread of how gender in agricultural research is depoliticized, the authors argue that without a strong conceptual and methodological foundation, intersectional studies in agriculture for research and development risk treating multiple identities as standalone 'tick box' variables, and not as multiple sources of marginalization. They use an empirical case study from the Ghanian livestock sector to demonstrate the application of intersectional analysis in AR4D based on a new conceptual framework and methodological approach. In conclusion the authors explore how AR4D can deepen its understanding of intersectionality and the potential application of this concept in a meaningful way that supports addressing multiple layers of inequalities and marginalization in agricultural research methods and practice.

Chapter 10 finds Cole and colleagues highlighting the important and unique characteristics that define feminist research approaches in agriculture, through the presentation of four purposively selected case studies. The case studies provide examples of how researchers working in agriculture can gradually adopt key feminist research principles. While conducting feminist research in agriculture is challenging and requires significant commitment to people and place, the authors argue that to transform agri-food systems to be more inclusive, equitable, and sustainable, feminist approaches must be used in all research in agriculture. The authors discuss how feminist research differs from gender research in in its aim to examine the diversity of women's experiences and how gender norms and power relations create inequalities between women and men and it's examination of why gender differences exist and challenges women's subordinate position while acknowledging the multiple variations between women that shape their experiences with oppression in different ways. The authors provide some good practices, while cautioning that there is often resistance to this king of research in the agriculture sector, due to the male dominance and bureaucratic structures in agriculture research for development.

A provocative concluding section in Chapter 11 sets the table for a reset, calling for the agricultural development community to: acknowledge the gendered hierarchies and power dynamics that are built into agriculture, recognize the interconnectedness of women's lives, bring women's rights out of the dark corner it has been relegated to in the sector, and reengineer patriarchal organizations and systems to address gender-based discrimination.

We hope you enjoy this book,

Jemimah Njuki, Hale Ann Tufan, Vivian Polar, Hugo Campos, Monifa Morgan-Bell and Vicki Wilde

References

Alsgaard H (2012) Rural inheritance: gender disparities in farm transmission. NDL Rev 88:347

Battilana S, Dhanraj D (Undated) Gender myths that instrumentalise women

Bhattacharya N (2019) The great agrarian conquest: the colonial reshaping of a rural world. SUNY Press, Albany

Carter A (2017) Placeholders and changemakers: women farmland owners navigating gendered expectations. Rural Sociol 82(3):499–523

Cernea MM (2005) Studying the culture of agriculture. Cult Agri 27(2):73–87

Chant S (2012) The disappearing of 'smart economics'? The world development report 2012 on gender equality: some concerns about the preparatory process and the prospects for paradigm change. Glob Soc Policy 12(2):198–218

Cornwall A (2018) Beyond "empowerment lite": Women's empowerment, neoliberal development and global justice. Cad Pagu 52:185202

Crenshaw K (1989) Demarginalizing the intersection of race and sex: a black feminist critique of antidiscrimination doctrine, feminist theory and antiracist politics. U Chi L Rev 1989(1):Article 8. http://chicagounbound.uchicago.edu/uclf/vol1989/iss1/8

Cullen B, Snyder KA, Rubin D, Tufan HA (2023) 'They think we are delaying their outputs'. The challenges of interdisciplinary research: understanding power dynamics between social and biophysical scientists in international crop breeding teams. Front in Sus Food Sys 7:1250709.

Echols A (1989) Daring to be bad: radical feminism in America, 1967–1975, vol 3. University of Minnesota Press, Minneapolis

Esteban-Guitart M (2016) Funds of identity: connecting meaningful learning experiences in and out of school. Cambridge University Press, New York

Evans-Winters VE, Esposito J (2019) Intersectionality in education research: methodology as critical inquiry and praxis. In: Qualitative inquiry at a crossroads. Routledge, New York, pp 52–64

FAO (2011) The state of food and agriculture 2010–11. Women in agriculture: closing the gender gap for development. FAO, Rome

FAO (2023) The status of women in agrifood systems. FAO, Rome. https://doi.org/10.4060/cc5343en

Farhall K, Rickards L (2021) The "gender agenda" in agriculture for development and its (lack of) alignment with feminist scholarship. Front Sustain Food Syst 5:573424. https://doi.org/10.3389/fsufs.2021.573424

Ferguson L (2015) "This is our gender person" the messy business of working as a gender expert in international development. Int Fem J Polit 17(3):380–397

Ferree MM, Tripp AM (eds) (2006) Global feminism: transnational women's activism, organizing, and human rights. NYU Press, New York

Gaard G (ed) (1993) Ecofeminism, vol 21. Temple University Press, Philadelphia

Gaard G (2017) Critical ecofeminism. Lexington Books, Lanham

Giddens A (2001) Sociology. Polity Press, Oxford

Gillborn D, Ladson-Billings G (2009) Education and critical race theory. In: The Routledge international handbook of the sociology of education. Routledge, London, pp 37–47

Hemmings C (2012) Affective solidarity: feminist reflexivity and political transformation. Fem Theory 13(2):147–161

Kabeer N (1995) Targeting women or transforming institutions? Dev Pract 5(2):108–116

Kiyama JM, Rios-Aguilar C (eds) (2017) Funds of knowledge in higher education: honoring students' cultural experiences and resources as strengths. Routledge, New York

Ladson-Billings G, Tate WF (1995) Toward a critical race theory of education. Teach Coll Rec 97(1):47–68

Lynn M, Dixson AD (eds) (2013) Handbook of critical race theory in education (pp. 181–194). New York, NY: Routledge.

Maynard M (1995) Beyond the 'big three': the development of feminist theory into the 1990s. Women's Hist Rev 4(3):259–281

Mehta B (2020) Jahaji-bahin feminism: a de-colonial Indo-Caribbean consciousness. South Asian Diaspora 12(2):179–194

Merrill-Sands D, Fletcher J, Acosta A, Andrews N, Harvey M (1999) Engendering organizational change: a case study of strengthening gender-equity and organizational effectiveness in an international agricultural research institute. In: Gender at work: Organizational change for equality. Kumarian Press, Inc., West Hartford, pp 77–128

Moser CON (1993) Gender Planning and Development: Theory, Practice, and Training. Routledge, London

Mukhopadhyay M (2014) Mainstreaming gender or reconstituting the mainstream? Gender knowledge in development. J Int Dev 26(3):356–367

Mukhopadhyay M, Wong F (2007) Revisiting gender training: the making and remaking of gender knowledge: a global sourcebook. KIT Publishers, Amsterdam

Njuki JE, Malapit S, Meinzen-Dick HJ, Bryan RS, Quisumbing E (2022) AR 2021. "A review of evidence on gender equality, women's empowerment, and food systems". Glob Food Sec 33:100622

Omi M, Winant H (2014) Racial formation in the United States. Routledge, New York

Ossome L (2022) Third world feminist agrarian struggles and the colonial question for transnational feminist solidarity. Agenda 36:18–28

Patel L (2014) Countering coloniality in educational research: from ownership to answerability. Educ Stud 50(4):357–377

Patton LD (2016) Disrupting postsecondary prose: toward a critical race theory of higher education. Urban Educ 51(3):315–342

Plumwood V (1986) Ecofeminism: an overview and discussion of positions and arguments. Australas J Philos 64(sup1):120–138. Gaard, Greta, ed. Ecofeminism. Vol. 21. Temple University Press, 1993

Robinson KJ (2020) Designing the legal architecture to protect education as a civil right. Ind LJ 96:51

Saugeres L (2002) Of tractors and men: masculinity, technology and power in a French farming community. Sociol Rural 42(2):143–159

Shisler, R. C., & Sbicca, J. (2019). Agriculture as carework: The contradictions of performing femininity in a maledominated occupation. Society Nat Res 32(8):875–892

Shiva V, Mies M (2014) Ecofeminism. Bloomsbury Publishing, London

Smith LT (2012) Decolonizing methodologies: research and indigenous peoples, 2nd edn. Zed Books, London

Solorzano DG (1997) Images and words that wound: critical race theory, racial stereotyping, and teacher education. Teach Educ Q 24:5–19

Thompson D (2001) Radical feminism today. Sage Publications, London

Tong R (2009) Feminist thought: a more comprehensive introduction. West View Press, University of North Carolina, Charlotte

World Bank (2012) World Development Report 2012: Gender Equality and Development. World Bank Group, Washington DC

Yosso TJ (2005) Whose culture has capital? A critical race theory discussion of community cultural wealth. Race Ethn Educ 8(1):69–91

Chapter 2
Let's Talk About Land: A Tale of Two Narratives in Gender and Agricultural Development

Hale Ann Tufan, Aubryn Sidle ⓘ, and Kendra Kintzi ⓘ

Abstract This chapter analyzes dominant narratives around gender and agricultural development and how they relate to critical feminist thought, between 1970–2020. We outline our heuristic approach by introducing the idea of the gender tree to frame the argument, and to summarize major strains in feminist theoretical frameworks. We present two historical narratives to illustrate liberal and critical approaches: *The undying allure of liberal feminism: Gender and agricultural development,* and *Critical approaches to land and gender justice.* We show how the mainstream narratives in agricultural development largely continued to address the consequences of gender inequality, while more critical approaches focused on land rights and access, challenging and targeting the root causes of gender-inequality. We bring attention to the need to put these two narratives more regularly in conversation, for mainstream gender and agriculture narratives to move closer to addressing the root causes of inequality through gender-transformative approaches, community interventions and a focus on gender norms and masculinities.

2.1 Introduction

For over half a century, women farmers around the world have been constructed as "objects" of development, and to a more limited degree, agents of agrarian change (Boserup 1970). Large multinational aid organizations often portray women as the backbone of rural economies, citing data that women comprise half of the agricultural labor force globally, that their farms are up to 30% less productive, and that if this gap were to be filled, this alone would lift 150 million people out of poverty (FAO 2011). Women are at the center of sustaining life and performing the labor of

H. A. Tufan (✉) · A. Sidle · K. Kintzi
Cornell University, Ithaca, NY, USA
e-mail: hat36@cornell.edu; as286@cornell.edu; kmk337@cornell.edu

© The Author(s) 2025
J. Njuki et al. (eds.), *Gender, Power and Politics in Agriculture,*
https://doi.org/10.1007/978-3-031-60986-2_2

15

social reproduction across the Global South.[1] How did women farmers come to be construed in this way, and how does the story of women's role in agricultural development illuminate broader trends and tensions within global development practice?

At face value, the history of gender and agricultural development reflects the influence of Western, liberal feminist thinking and the ascendance of neo-classical economics and growth-based approaches to development. Yet this characterization of history is only partially complete, obscuring a robust body of work in critical theories of feminism and development and the material implications of these theories for practice. A more comprehensive view of the history of gender and agricultural development can be seen by foregrounding questions of land and land tenure, helping us unearth the ways that women are constructed in theory and histories and the way land is positioned as either a productive asset or as the site of social reproduction and social change.

In this chapter, we examine these two streams of thought—liberal and critical—through the allegory of a tree, exploring how the mainstream gender and agricultural development literature remains focused on measurable productivity gains driven by a continuing legacy of liberal feminism through the narrative *"The undying allure of liberal feminism: Gender and agricultural development,"* while land justice literature evolved to explore root causes of inequality through a critical feminist and decolonial lens through the narrative *"Critical approaches to land and gender justice"*. We then discuss points of convergence that are emerging in contemporary literature, focusing on gender-transformative approaches, and outline key promising practices for incorporating decolonial, equity, and justice perspectives into gender and agricultural development practice.

We write this chapter from a position of privilege and power. Our perspective is informed by our experiences working at the intersection of critical development theory and practice. Collectively, we have combined experience conducting qualitative and quantitative research in countries across sub-Saharan Africa, Southwest Asia/Middle East and North Africa, and South Asia. Our approach to this article is rooted in our combined experience working in the field of gender and development, and by our collective ethical and political commitments to gender justice, social equity, and decolonization. At the same time, we recognize our privileged positionality as white and white-presenting women working within a well-resourced institutional context in the Global North. We are particularly sensitive to, and critical of, the colonial and imperial histories that have privileged the voices and perspectives of white women in shaping the evolution of feminist movements (Zakaria 2021). In this article, we intentionally center decolonial approaches by refusing to treat global women as a homogenous category (Mohanty 1984), and we seek to incorporate key insights from critical theory as pathways towards liberatory practice (hooks 1991).

---

[1] We draw on the work of Sud and Sánchez-Ancochea, who theorize the term "Global South" as "a territorial, relational, structural and political construct" that indexes ways of interrogating global distributions of power (p.1). Our use of the term in this paper is designed to call attention to uneven power relations at a global scale, in order to investigate possible pathways for liberatory practice (Sud and Sánchez-Ancochea 2022).

In the following section we give a brief background on critical development scholarship and review key feminist theoretical frameworks, and how they view the problem of gender inequality through a historical lens (Fig. 2.1). We explain how we frame our argument through the gender tree (Fig. 2.2), in reference to gender analysis approaches that differentiate root causes of gender inequality and discrimination, from their manifested consequences (Oxfam 2014).

Examining the Root Causes and Consequences of Gender Inequality in Agricultural Development   Although the rise of liberal feminism and neo-classical economics have dominated gender and agricultural development spaces over the past 50 years, there are equally well-developed conversations in critical theory and feminisms that have important implications for the future of gender and agriculture practice. These two narratives treat development in radically different ways, as either a political process of social reproduction (critical approaches), or an apolitical march towards modernity (liberal approaches). Both views focus on the importance of land to women and agriculture, but for different ends: critical feminist narratives center land as a right and the basis of empowerment, while liberal feminist approaches treat land as an essential asset that is necessary for agricultural and economic productivity.

Although land is a central issue in both views, there is little cross referencing between feminist rural sociology (the literature historically concerned with critical approaches to land and agriculture), the gender and agricultural development publications, and the official documentation from development agencies. This lack of integration is explained by the liberal feminist focus on alleviating the symptoms of gender inequality while sidelining the root cause of the illness: historical and ongoing land dispossession and exploitation. Critical and liberal narratives reflect two divergent interpretations of the history of gender and agriculture. Mainstream

Fig. 2.1 Major theoretical strains in feminist theory

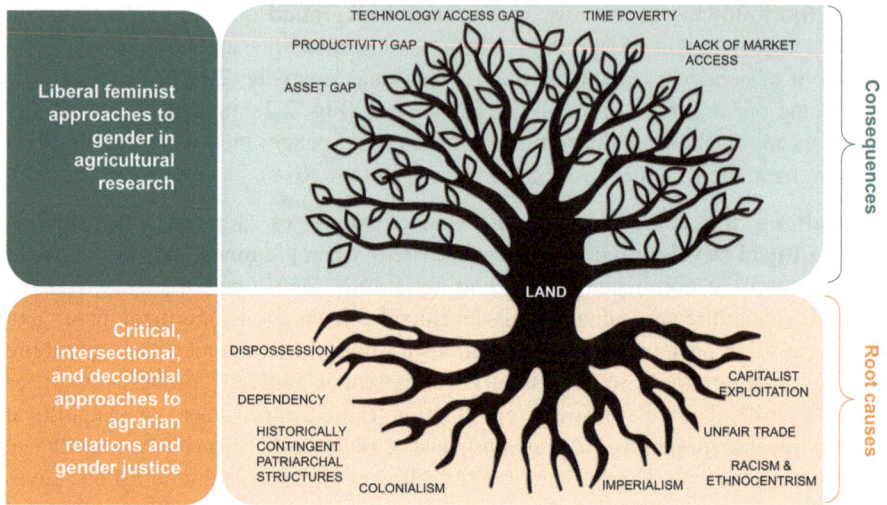

Fig. 2.2 Relating the two narratives through an allegory of a gender tree

approaches to gender and agricultural development tend towards investigating "fixes" to the empirical manifestations of gender inequality, and in doing so fail to "see" and therefore alleviate the underlying root causes producing inequality. In contrast, critical approaches to gender and land seek to understand colonial and neocolonial processes and root causes of the structural inequality faced by rural women and farming communities. We summarize these two narratives and their relationship in the allegory of a gender tree—where land represents the trunk of the tree and the core of both narratives, and the branches and roots are key focuses of each narrative's explanation for gender inequality (Fig. 2.2).

In the next section, we explore two narratives in the gender and agricultural development literature that reveal divergent orientations to the question of how to achieve gender equality, land justice, and agrarian development. We name these narratives *The undying allure of liberal feminism: Gender and agricultural development* for liberal approaches focused on "consequences", and *Critical approaches to land and gender justice* for critical approaches focused on root causes. This discussion traces these approaches through the distinct fields of gender and agricultural development, critical agrarian studies, and feminist political ecology literature. We provide a chronological overview to narrative one, which focuses on the undying allure of liberal economic approaches to enhancing women's role in agricultural development. With narrative two, our approach is not explicitly chronological, but illustrates the body of critique that has existed alongside and underneath mainstream development paradigms. We examine the intellectual roots of these critiques, how they speak to ongoing tensions within the world of gender and agricultural development, and where historically, more critical viewpoints have bubbled up to powerfully impact mainstream practices.

2.2   A Tale of Two Narratives: Root Causes and Symptomatic Branches of Gender Inequity

2.2.1   The Undying Allure of Liberal Feminism: Gender in Agricultural Research for Development

By most accounts, the history of gender and development began with Esther Boserup. In her seminal book *Woman's Role in Economic Development,* originally published in 1970, she highlights "female farming systems" in sub-Saharan Africa to demonstrate women's roles in the rural economy that had been invisible in mainstream, liberal economic development approaches. Boserup used this observation to argue that women had been excluded from economic opportunities, and development efforts, which ushered in the dawn of liberal feminism and gender in development history. It is revealing to superimpose Boserup's initial liberal economic worldviews from 50 years ago with contemporary arguments around agricultural productivity and economic growth that abound in agricultural research for development. For example, Okali (2012) speaks of these dominant paradigms of women as "cardboard victims or heroines," drawing out key narratives as political discourses that describe women as altruistic, risk-averse, ignored by service providers, lacking agency in intrahousehold power relations, and lacking control of their labor and resources, yet at the same time feted as the heroines for rural economic empowerment (see also Whitehead 2000). While the language around gender equality and women's empowerment has successfully permeated the agricultural research for development discourse, it remains largely detached from higher order principles of feminist engagement (Cornwall and Rivas 2015). Current agricultural research for development paradigms still instrumentalize women for development gains (Farhall and Rickards 2021). In the following section, we give a broad summary of the progression of narratives around gender and agricultural research for development. This is not intended as a comprehensive historical analysis, rather it is a synthetic overview to compare and contrast with critical scholarship. For simplicity, we have presented the historical narrative as a linear temporal progression, while in reality we see influences of different feminist theoretical frameworks (Fig. 2.1) piercing through and influencing agricultural development practice across this timeline.

2.2.1.1   Awakening and Diagnosing "Problems" (1970s and 1980s)

In the 1970s and 1980s, the Women in Development (WID) paradigms that grew out of liberal feminism and Ester Boserup's early work yielded policies that focused on the productive capacities of women as primary food producers, and the drivers of population growth and fertility transitions. Under the framework of efficiency, women were seen as important producers as they provided a source of labor to increase agricultural production, while men remained *de facto* heads of households and directed the farm. Men were portrayed as market-oriented and using modern farming technology, while women were portrayed as subsistence producers for household food security (Okali 2012), which has remained a constant narrative in the ensuing 50 years. This early characterization of household headship has also endured in agricultural economics literature, where comparisons of female- and male-headed households continue to create false dichotomies of measurable productivity gaps to be filled by agricultural development interventions, and do not account for the role of women in decision making or control of assets in male-headed households (Doss 2013; Deere 2010).

Midway through the 1970s, farming systems thinking emerged as an important paradigm that led to the "re-peopling of agriculture" (van der Burg 2020). As farming systems and farmer first thinking shaped agricultural research for development, focus shifted to capturing the demands of resource poor and marginalized farmers. Gender was mostly treated as an analytical category, with little attention paid to gender relations. Notably though, the Rockefeller Foundation supported the integration of social scientists to support international agricultural researchers to unpack this "re-peopling" of agriculture with a gender lens (van der Burg 2020). This integration of systems thinking into agricultural research was part of a broader expansion of systems thinking as a positivist, interdisciplinary approach to solving complex social challenges using tactical methods that often lacked robust theoretical engagement (Carr-Chellman and Carr-Chellman 2020).

In the 1980s, critiques of modernization theory (including Marxist feminist ones), led to sex-disaggregated data collection as the primary mode of diagnosing women's inclusion, while retaining a primary focus on economic development. Through the 1980s major donors commissioned reports and studies to understand "gender issues" in agricultural development (e.g., Overholt et al. 1985), rife with recommendations on interventions to raise women's productivity. This signified a shift in agricultural development narratives from awakening to problem identification (diagnosis). The 1980s saw a sizable amount of emergent research literature around gender in agricultural research for development. The arrival of the "Harvard framework" (Overholt et al. 1985) opened up space for gender analysis to be mainstreamed into agricultural research for development. The framework's focus on gender roles in farming households led to its use as a widespread data collection approach. Yet its use led only to seeing manifested gender roles without exploring the underlying gender and power relations that caused them, and remained largely detached from social relations-focused approaches proposed by Moser (1993) around the same time. This focus on roles over relations is still noticeably present in current literature. The mid-1980s also saw the emergence of early literature that highlighted the unintended consequences of gender-blind development interventions (e.g., Dey 1985), gender analysis frameworks developed with a focus on agriculture (e.g., Feldstein and Poats 1989), the first documented conference on women and agricultural research for development (1983, organized by the International Rice Research Institute) and a first review on gender-related impacts of the CGIAR system (Jiggins 1986).

2.2.1.2 Promoting Participation (1990s)

Landmark shifts in gender in development thinking marked the period of 1990–2000. Naila Kabeer's critique of development planning (1991), together with influences from intersectional and decolonial feminist thinking on the importance of gender relations and the intersection of gender, class and race (DAWN 1986), shaped a Gender and Development (GAD) landscape that gave rise to a focus on women's empowerment. Gender mainstreaming and participatory approaches to women's inclusion emerge in this era as strategies for achieving the institutional power shifts suggested by the "empowerment" concept. Unfortunately, while these efforts made some headway, they largely failed to accomplish the intended transformation as gender mainstreaming and participatory processes were often tokenized by broader institutional and social dynamics which served to sideline the intended power-shifts, and very few interventions actually targeted underlying gender relations that served to construct gender inequality as per the GAD view (Moser and Moser 2005; Rao and Kelleher 2005).

In the 1990s, the viability of structural adjustment programs to achieve their goals was being called into question in the agricultural development context (Saito et al. 1994). Kingiri (2013) describes this period as women being spectators or innocent victims rather than players in adjustment programing, due to the lack of attention to women's access to and control over resources. Large organizations such as the Food and Agriculture Organization (FAO) rolled out their first women in development plan of action (1989–1995) focusing on improving women's access to resources and training. These larger trends of an initial focus on access to resources and participation paved the way for an expansion of diagnostic studies using approaches with a focus on providing assets and means to increase productivity. This focus on sex-disaggregated quantitative data collection to inform policy was formalized with the inception of the IFPRI Gender and Intrahousehold Dynamics program (1992–2003), while a continuation of reductionist focus on female-headed households continued more broadly, avoiding discussions around power relations through an excessive focus on feminization of poverty and women as smallholder heroines (Geisler 1993).

We see a reflection of these broader trends in agricultural research for development, as role and asset analysis started to enter the realm of agricultural technology development. Yet these efforts were still marred by systematic gender bias in the agricultural research system, where voices of technologically tuned-in male household heads are deemed more valuable, as documented by Anne Ferguson (1994). By the mid-1990s farming systems research had started declining, while participatory research and gender analysis gained momentum, especially in CGIAR centers where "all centers are doing something" (Feldstein 1998). Feldstein meticulously documented an inventory of gender-related research and training in the CGIAR. The formalization of this focus on participatory research is embodied in the formation of the Participatory Research and Gender Analysis Program of the CGIAR in 1992.

The focus on participatory research, roles and asset analysis helped to circumvent the more challenging ideas that came with the rise of GAD. The participatory agricultural research agenda created space for claims around including women in agricultural research, helping programs avoid the critique of women being unintentional victims of technical changes (Okali 2012). These efforts remained mainly instrumental, ignoring gender relations and how these shaped the diagnostic results, while their claims of inclusion helped to diffuse political pressure and appealed to critical scholars. The mainstream trends of participatory research co-existed with seminal works by feminist scholars (see Sachs 1996) that brought much needed critical insight to the agricultural research for development ecosystem.

2.2.1.3   Minding the Gaps (2000s)

At the turn of the century, the donor community turned to private-sector involvement, driven by the hollowing out of public funding, the rise of philanthro-capitalism from the tech sector, and a broader shift towards privatization, bringing in concepts of results-based management and accountability to agricultural research for development. Framings around "meeting the target" made private-sector involvement seem sensible, while solving systematic and global inequality seem impractical and ideological (Fukuda-Parr 2016). The birth of the Millenium Development Goals (and later the Sustainable Development Goals) demanded a focus on measurable indicators for evidence-based decision making. Liberal feminism continued to drive a depoliticized gender agenda, with technocratic fixes, measurement and a focus on the bottom-line productivity which still characterizes the language and policies of large multinational organizations today. For example, the FAO Gender and Development Plan of Action (2002–2007) focused on emerging challenges such as the impact of globalization, population dynamics and pressure on natural resources, implemented through capacity development, raising awareness, and gender-sensitive indicators and statistics (FAO 2011). At the time of randomized control trials as the measure of successful technology impact evaluations, together with a neoliberal emphasis on the importance of markets and value chains, the dominant narratives in gender and agricultural research for development focused on gender gaps and the resulting "lost opportunity" of rural economic development, shying away from seemingly impractical and ideological narratives around rights and gender justice.

   This gap era is defined by popular victim/heroine narratives where women carry the labor burden of smallholder farming, are poorer, and embrace farming for food security typified by quotes such as: "Rural women are the main producers of the world's staple crops—rice, wheat, and maize—which provide up to 90 percent of the food consumed by the rural poor. Women sow, weed, apply fertilizer and pesticides, and harvest and thresh crops" (World Bank, FAO, and IFAD 2009, 522). This victim narrative is then complemented with the heroine, highlighting that closing the asset and resource gap will lead to sizeable reductions in poverty levels, more income and more productivity. A hyper-focus on asset gaps led to major investments by donors, for example the Gender, Agriculture and Asset Project by IFPRI. Gender inequality in asset access and control came to define the productivity gap, offering a convenient conduit for technocratic fixes in agricultural research for development, and shifted the focus on access to "solutions". The gender agenda grew exponentially in prominence at the end of the 2000s with the publication of the "Gender in Agriculture Sourcebook" (World Bank, FAO and IFAD 2009), which offered reflections of gender in 16 different agricultural development topics, complete with thematic introductions, succinct guidelines for implementation and examples of innovative projects. While this important publication fell short of systematic and critical analysis of gender in agricultural research for development, it irrevocably cemented the importance of gender and agricultural development.

2.2.1.4 Reaching, Benefitting and Empowering (2010s)

The dominant productivity and asset gap narrative in gender in agriculture in the 2000s, which signified a return to liberal feminism, continued strongly into the next decade (Quisumbing et al. 2014; FAO 2011). At the same time, as mainstream approaches absorbed limited parts of the influence of decolonial and Marxist feminist critiques, we start to see an agenda that goes beyond productionist arguments. The systematic documentation of asset and productivity gaps in the "gap era" combined with the resurgence of attention to concepts of empowerment, created a need for more standardized and measurable framings of women's empowerment in agriculture.

The publication of the first paper on Women's Empowerment in Agriculture Index (WEIA) (Alkire et al. 2013) ignited strong investment in and attention to WEIA, and its subsequent derivatives over the next decade (see Malapit et al. 2019 for a review). Fruits of the Gender Agriculture and Assets Project (GAAP) documenting interventions related to use, access and control of agricultural assets examined underlying household and community attitudes about women's work, participation in decision making (Johnson et al. 2016) and led to the development of the "reach-benefit-empower" framework (Theis and Meinzen-Dick 2016). Here we also see the selective incorporation of critical and decolonial feminist critiques into the liberal feminist traditions of thought, which converge in the form of Women's Economic Empowerment (WEE) as a solution to address the productivity and asset gap, and a focus on interventions that "level the playing field". Women's economic empowerment, particularly in the Global South was also linked to their "technological empowerment" (Kingiri 2013) through greater access to new agricultural technologies.

In this decade scholarship from feminist economists brought much needed critiques to neo-classical economics and its treatment of agriculture and development. We see a convergence here in scholarship with critical approaches to land and gender justice (summarized below), through the particular attention feminist economists paid to household behavior (Doss and Quisumbing 2021), land access and land rights. Examining land deals, Doss et al. (2014) reveal how these link to broader dispossession and draws attention to the knowledge gap on gender and land rights. Focusing on women's land rights as a pathway to poverty reduction, Meinzen-Dick et al. (2017) revealed that "Many gaps in the evidence arise from a failure to account for the complexity of land rights regimes, the measurement of land rights at the household level, the lack of attention paid to gender roles, and the lack of studies from countries outside Africa." (p v). With more contemporary research and conceptual attention to land rights and land access (see Doss and Meinzen-Dick 2020; Slavchevska et al. 2021), it is becoming more "mainstream" that land is more than just another asset, but a central tenet that binds together the root causes and consequences of gender injustice in agricultural development.

2.2.2 Narrative 2- Critical Approaches to Land and Gender Justice

Alongside the trajectory of mainstream liberal approaches to gender in agricultural development, a flourishing body of critical scholarship on land justice has punctuated, and at times powerfully reshaped, mainstream approaches to the question of gender in agrarian change. These critiques emerged across a variety of disciplines and geographic contexts, drawing together threads from decolonial, intersectional feminist, and critical agrarian studies approaches to offer insights into how gender relations in the (post)colonial world came to be the way they are, and how they might be different. Here, we map out three dimensions of these critiques and how they challenge key presuppositions within the prevailing liberal economic framework for advancing gender equity in agricultural development.

2.2.2.1   Insights from Critical Development Scholarship

Since the crystallization of "Development" in the post-World War II era, an increasingly extensive and far-reaching bureaucracy of private, national, and multilateral development actors has emerged as an intentional, coordinated project of international intervention to correct for the disruptive and destabilizing effects of capitalist expansion (Hart 2001). While the places and programs gathered through this bureaucratic apparatus are diverse, postwar Development practices are joined together by a core focus on poverty alleviation, modernization, and social progress that is often articulated through an epistemic paradigm of liberal economic growth. Within this liberal paradigm, practitioners employ discursive practices that frame Development as a neutral project of technical intervention, depoliticizing the problems of poverty and inequality in a way that "renders invisible the power relations that produce it" (Ferguson 1990 p. 256). However, as many critical development scholars have shown, attempts to render Development technical and apolitical is neither secure nor complete (Li 2007). These critiques help us view contemporary gender and agricultural development projects historically and politically, highlighting the ways that these projects are enrolled in broader discursive and material

Development practices. While mainstream approaches to gender and agricultural development often employ technocratic, depoliticized language, the effects of these programs often exceed their intended parameters (Adely 2004). Whether intentional or not, agricultural development projects shape how land is accessed, used, and cared for, sometimes leading to more gender equitable outcomes and sometimes entrenching gendered forms of exclusion and inequality (Hajjar et al. 2020). Understanding the contemporary landscape of gender and agricultural development practice requires critical analysis of how particular meanings, narratives, and approaches became dominant, and how these practices continue to be critiqued and contested across the world today. It also requires recognizing the agency of women beneficiaries enrolled in these programs, and how they envision and enact gender justice on the land.

2.2.2.2 Reframing the Rural: Insights from Critical Agrarian Studies

While most mainstream liberal approaches to gender and agriculture target inter-ventions *on* rural land, few critically engage the question of land itself. Land tenure and the political economy of land distribution is most commonly treated as a given, and land itself as the platform for the development of robust private property regimes, neglecting both situated histories of communal tenure and enduring ques-tions of land reform (Manji 2006). Yet scholarship in the critical agrarian studies tradition has long called attention to the multiple, contested meanings of land itself—and to the essential role of land relations in shaping the kinds of social, envi-ronmental, and political development that become possible (Wolford 2010). Scholarship in this tradition burrows into the subsoil of land: the histories of how it came to be valued in specific ways, the social relations that shape how it is used and the multiple meanings it holds, and the role that it plays in supporting or maintain-ing particular power relations and political-economic structures (Carney and Watts 1990; Faxon 2020). Bringing focus to the subterranean roots of agricultural produc-tion, scholarship in critical agrarian studies positions land as contested terrain and the key site of struggle over the economic, social, and political impacts of develop-ment efforts (Li 2014). Land and land tenure are key to sustaining not just agricul-tural output, but the livelihoods, lifeways, and broader social relations that organize society in agrarian contexts. Within the critical agrarian studies tradition, three key questions continue to play a powerful role in shaping the gendered effects of global agricultural development efforts: (1) how land is used (2) how land is distributed, and (3) how life on the land is reproduced.

How Is Land Used? The first question, how land is used, invites us to consider both the socioeconomic and environmental implications of particular modes of pro-duction on land. Critical agrarian studies scholarship examines how distinct agricul-tural practices play a constitutive role in producing and maintaining larger structures of systemic inequality (Bernstein and Byres 2001). While scholarship and practice oriented around the liberal economic worldview continues to prioritize increased productivity, higher yields, and connecting farms to markets, critical agrarian scholarship has repeatedly shown how linking smallholder farmers to international agricultural markets introduces new forms of risk and precarity, while shifting the material foundation of agrarian societies to become increasingly dependent upon unstable global commodity market dynamics (Akram-Lodhi and Kay 2010; Goodman and Watts 1997).

This body of scholarship recasts smallholder, subsistence agriculture and agroecological approaches as key strategies for maintaining agrarian lifeways and supporting more resilient, independent forms of agricultural production (Altieri and Toledo 2011). This work has also called important attention to the fact that smallholder, subsistence agriculture in the Global South is predominantly labored by women, highlighting the ways that agricultural production is intimately linked to gendered relations of production (Bezner Kerr et al. 2019). Critical agrarian studies offers an important critique of production-oriented agricultural development efforts, shedding light on the environmental impacts of such efforts that consolidate land, intensify natural resource extraction, and increase inputs. This creates a cascading effect of increasing dependence that depletes the capacity of soils and farmland to sustain biodiverse forms of life, the impacts of which are disproportionately felt by women and children (Hajjar et al. 2020). Work in feminist political ecology and agroecology has called for a critical reevaluation of the pivotal role of small-scale, household and community-focused agricultural practices as a powerful alternative pathway towards sustainable, inclusive development. This critical scholarship shows how agricultural development projects disproportionately impact women, who comprise a significant proportion of the population of smallholder and subsistence farmers in the Global South and because the effects of how land is used are felt first by women and their dependents.

How Is Land Distributed? The question of how land is distributed matters for gender inequality because it helps us analyze broader patterns of inequality regarding who owns and controls land rights. Analyzing how land is distributed and patterns of ownership (tenure) and use (usufruct rights), helps us see how agricultural production on the land supports broader patterns of political and economic control, and pinpoints the central role of land ownership in shaping the socioeconomic and political outcomes of development efforts (Brenner 1976). Critical agrarian scholars have shown how specific multilateral agricultural development policies focusing on the promotion of primary commodity exports actually undermine macroeconomic stability at regional and national scales. Critical work on key crops like bananas, sugarcane, and cacao has repeatedly shown both the ecological destructiveness of monoculture cultivation and the adverse economic and sociopolitical impacts of channeling significant portions of regional and national economies into dependence on single, export-oriented agricultural commodities (Alarcón 2022; Taylor 2016). Furthermore, the multilateral promotion of export-oriented agricultural commodities as foundational development practice contributes to the furthering of inequality at global scales, as the commodification of key crops enables the devaluation of the labor of primary production and the concentration of profit by transnational corporate entities (Mintz 1986).

Finally, critical scholarship in this vein has repeatedly shown how uneven land ownership shapes whether agricultural development contributes to broad-based, inclusive gains, or exacerbates structural inequality and the capture of key benefits by a narrow cadre of landed elites (Hall 2013). Activists and critical scholars around the world have long called for an emphasis of land redistribution and land justice, and empirical examinations of land redistribution in South Korea and Taiwan reveal the powerful impact of the equitable distribution of land in democratizing development outcomes in the agricultural sector (Walinsky and Ladejinsky 1993). Key contributions of feminist scholars continue to document how the uneven distribution of land has acutely gendered effects (Hajjar et al. 2020; Benería and Sen 1981). Historical patterns of land ownership and use have been gendered and disproportionately favorable to men. Land reform constituted a central dimension of (post) colonial mobilization across much of Latin America and sub-Saharan Africa, yet the unfinished business of land reform continues to drive uneven development outcomes in many countries to this day. As noted above in Sect. 2.2.2.1, there is some convergence between this scholarship and recent work in feminist economics that critiques neo-classical approaches to agricultural development and its treatment of questions of growth and inequality (Sent and van Staveren 2019).

How Is Life on the Land Reproduced? Finally, the question of how life on the land is reproduced brings a focus to the gendered relations of production and social reproduction in agrarian contexts in the Global South. Marxist feminist scholarship has long emphasized the crucial importance of moving beyond a simplified view of households as units of economic production, to see how political dynamics of power and inequality are reproduced, and contested, at the household scale (Benería and Sen 1981). This diverse body of scholarship highlights women's role in social reproduction, drawing out the constitutive role of women's unpaid labor in the home and in the field. Moreover, critical feminist scholarship denaturalizes the feminization of poverty, prioritizing and politicizing the questions of how it came to be that women-headed households engaged in subsistence farming predominate across the Global South (Hart 2001). This work calls us to rethink key connections between land, labor, and livelihoods, showing how women's work in agricultural subsistence cannot be delinked from broader questions of labor outmigration, land ownership and dispossession, and the struggle to make a life in the wake of colonial and imperial resource extraction.

2.2.2.3 Decolonizing Women's Empowerment: From Land Rights to Intersectional Justice

Insights from critical agrarian studies and Marxist feminism are expanded upon by decolonial feminist discourse which situates gender inequality in capitalist histories of oppression beginning with colonial and neo-colonial processes of dispossession and hegemony (Anzaldúa 1987; Mohanty 1984; Sen and Grown 1987). Thus women's alienation from land is a product of histories of racialized exploitation, perpetuated by current policies, discourse, and ideologies that present Western ideals, ideas and societies as the dominant standard (Mohanty 1984). Decolonial and Third World feminist theorists direct their critiques of mainstream gender and development policies towards modernization theory and the erasure of colonial processes which produced structural inequalities, and the neocolonial processes which continue to reproduce these inequalities through current social, political and economic discourse. A particular point of contention for decolonial scholars is the ways in which the ideal of the "empowered" woman in development policy is essentially a replica of the independent, individualistic and autonomous female ideal produced by white liberal feminists (Abu-Lughod 1998). Since women's empowerment has been an important point of policy in gender and development, and in gender and agriculture research, this critique warrants further examination.

Empowerment in some of its earliest iterations by decolonial feminists refers to a *process* of political mobilization in which women build political and social movements to transform the balance of power (Sen and Grown 1987; Stromquist 1999). This original definition is in strict opposition to more commonly deployed conceptualizations of WEE which view women's empowerment as an instrument for other development goals (such as food security or economic growth) and which places empowerment within an economic efficiency view of women as rational economic actors (Sardenberg 2008, pp. 18–19). Naila Kabeer has famously defined women's empowerment as agency or the "power to" combined with resources needed to allow people to live a life of their own choosing (Kabeer 1999, p. 438). Implicit in Kabeer's definition is that these so-called "liberal" definitions of empowerment (e.g., access to land, resources, capital) are included within, but not inclusive of "liberating empowerment" (Sardenberg 2008), which would necessitate women's agency and related structural shifts.

Decolonial feminist theory thus demands a move beyond binary paradigms, arguing that transforming gender power relations by simply turning patriarchy on its head through the economic empowerment of women, merely reproduces oppressive structures (Sandoval 1991; Mohanty 1984). For gender and development practitioners, this is perhaps one of the most useful points of critique: that practice, policy and research should move away from the reproduction of social inequality (regardless of "who" holds power) and instead, should conceptualize new forms of social justice outside of the organizing frame of patriarchy (Kandiyoti 1998).

We see elements of decolonial thinking in the current work being done in gender and agriculture on intersectionality. Intersectional feminists have long called attention to how gender and race (Crenshaw 1990), gender and class (Sen and Grown 1987), and gender and sexual orientation and disability (Collins et al. 2021; Majeedullah et al. 2016), shape women's experiences. Recent work by Leder and Sachs (2019) and others, shows how mainstream approaches to measuring women's empowerment actually obscure material differences in how women experience "empowerment," based on differences in caste, class and inter-household power dynamics. Whereas intersectionality is a powerful tool for uncovering structural inequality and for explaining why interventions fail, decolonial theory provides both methodological and practical tools for conceptualizing social justice and "transformative" approaches.

The diverse body of critical scholarship on gender and agricultural development helps us see how questions of land and land tenure profoundly shape the trajectories of agricultural development and the possibilities for equitable, broad-based, gender-inclusive development. The questions of how land is used and distributed help us analyze patterns of land use, and tenure, in order to investigate how particular projects of agricultural intervention might deliver equitable gains in rural landscapes or further entrench power differences between large landowners and smallholder farmers. These questions are particularly important in the Global South, where women comprise a large proportion of smallholder farmers and where land rights, land tenure, and different forms of agricultural production are often highly gendered. Finally, the question of how life on the land is reproduced provides an entry point to better understand and center women's voices in agricultural development efforts. In particular, decolonial and intersectional approaches help us shift from diagnosing inequality to centering just, transformative approaches.

2.3 Let's Talk About Land: Centering Land Rights and Land Use in Agricultural Development Practice

We framed this piece as a tale of two narratives, feeding off of observed dichotomies and divergences in the gender and agriculture mainstream literature, and critical feminist agrarian literature, as a heuristic approach to describe them in relation to one another. While there is ample heterogeneity within these narratives, here we draw out key ideas and practices that have shaped the evolution of gender and agricultural development.

Focusing on the branches of the tree, approaches within the mainstream, liberal paradigm of gender and agricultural research primarily engage with, measure, and focus on "solving" the symptoms of gender inequality in agrarian contexts. For example, there is a significant focus in contemporary agricultural development practice on women's lack of access to markets and assets and resource gaps. This approach problematizes the lack of access to markets and assets as a solvable problem, a framing that simplifies complex social, historical and political contexts into observable, measurable and visible gaps and shortcomings. This plays well to donors focusing on metrics, impact evaluations and sustainable development goals, while circumventing serious critical self-reflections by individual researchers and development organizations by externalizing gender inequality. A focus on measurable consequences rather than causes also plays well to agricultural researchers, offering a simple way to understand and respond to gender issues by filling gaps with more interventions. This superficial/above ground thinking is also perpetuated through gender training programs that simplify complex theoretical concepts and processes of social change (Mukhopadhyay and Wong 2007; Tufan et al. 2021).

The roots of the tree, in contrast, represent structural causes and drivers that give rise to the symptoms, or expressions of gender inequality that are visible as the branches of the tree. The structural causes are rooted and complex, and emerge through historically specific patterns of inequity and exclusion that are linked to colonial and capitalist processes of extraction, accumulation, and dispossession. For example, a critical approach to the root causes of asset and resource gaps would consider colonial histories of land expropriation and contemporary patterns of agricultural consolidation. This approach would focus on the social relations that created uneven access to, and control over, assets and resources. Colonial histories and networks of economic integration link disparate geographic contexts around the world; however, root structures are place-based and difficult to scale, as structural drivers of inequality are intimately linked to power relations within particular places. As a result, identifying and analyzing root structures requires significant investments of time, resources, and modes of engagement that are more complex than the diagnostic work of mapping the symptoms of the branches. In addition, these root causes are inherently political in nature.

Binding the branches to the roots is the trunk of the tree, which represents critical feminist scholarship around land rights and gender justice described in the second narrative, as a connective bridge that helps ground the visible and measurable dominant narrative in agricultural development. Narratives around land rights help bring these two tales together, as without land there can be no agriculture. Land is a consistent part of the asset gap narrative strongly present in the branches, subject to myths around gender in agriculture (Doss et al. 2018) and yet access to land and land rights are inherently contextual, political, contested, and subject to relationships of power and patriarchy.

This linkage between consequence and root cause is gaining prominence in mainstream gender and agriculture narratives, signaling a promising move towards a more critical research agenda. Exciting new developments in gender and agricultural research draw on decolonial and intersectional feminism, and demonstrate a move back towards the critical feminist roots of gender and development through research into systems of power, patriarchy and gender norms (see chapters in Sachs 2019 and Sachs et al. 2020). This has gained prominence in mainstream agricultural research and development spaces as well, for example as researchers working in or with the CGIAR publish on these topics: focusing on perceptions of agricultural innovation by women in patriarchal family structures (Kawarazuka and Prain 2019), to the application of intersectionality to agricultural research design (Tavenner et al. 2022), critical feminist analysis of "women's empowerment" in agriculture (Tavenner and Crane 2022), critical analysis of the "gender agenda" in agricultural research (Farhall and Rickards 2021), gender norms and agency in agricultural innovation systems (Badstue et al. 2020), gender-transformative seed systems (Puskur et al. 2021), "feminist critical agronomy" framings to critically examine concepts of markets and demand in crop varietal development (Tarjem et al. 2023), and flipping the paradigm on gender equality as an outcome rather as an afterthought of agricultural research (see chapters in Pyburn and van Eerdewijk 2021).

Perhaps most notably, attention to gender-transformative approaches, norms, intersectionality and critical analysis is entering the lexicon of large multi-national donors such as The International Fund for Agricultural Development (IFAD), FAO, the World Food Programme (WFP) and the European Union (EU) (see for example JP GTA 2022; FAO 2023), signaling a significant shift in the narrative of agricultural research for development. This convergence of agendas offers a promising path forward, as the trajectory of gender in agricultural development enters a new era, moving away from a focus on the consequences of gender inequality (branches) towards rooting these debates in research and scholarship that unveils its root causes.

In this new era, there is a unique opportunity for researchers and practitioners to more systematically link branches to roots, or to "embrace the trunk" by centering discussions around land rights and use. Here, the gender tree offers an opportunity to seed discussions between critical scholars, researchers and practitioners. We present this heuristic as a starting point for dialogue and debate, with the goal of generating productive pathways that draw upon the full scope of theoretical work to bring us to more gender-just and equitable land futures. Land relations are core to each of the problems in gender and agricultural development that contemporary practitioners are trying to fix, with land access, tenure, usufruct systems and cultivation practices all playing a central role in creating and sustaining gendered, inequitable agricultural systems. Figure 2.3 demonstrates what this heuristic could look like for crop varietal development.

For many practitioners, it may be conceptually easier to address gender questions by mapping uneven access to technologies, inputs, and markets (branches), than to think about how systems of colonial expropriation or capitalist extraction have led to these outcomes (roots). Centering the discussion around land rights and use helps to illuminate the contextual political and patriarchal power dynamics that shape these land dynamics. Here we use land as an analytical entry point, opening discussions around how and why systemic root inequalities and injustices feed into

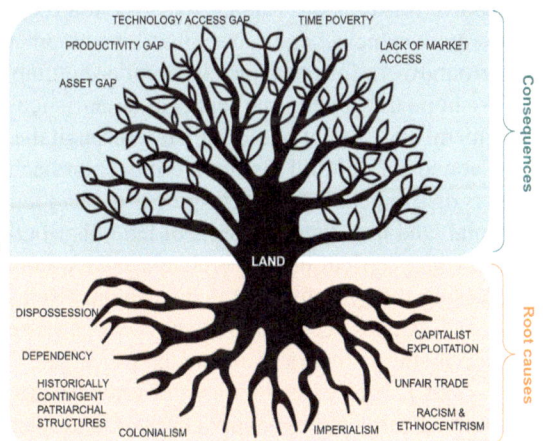

Observe consequences: Women have lower yields with the crop, spend more time on the plot, lack access to new varieties, inputs and markets

Understand land dynamics: What is known about women's land rights and use? How does this shape the time they spend on the plot, decisions made on varieties and markets? Is the land women have access to less fertile or further from the home?

Unearth root causes: Patriarchal land tenure limits women's rights and access to land, women are dispossessed from fertile plots, markets for this crop are increasingly consolidated to serve urban markets

Fig. 2.3 Let's talk about land: an example for crop varietal development

inequitable agricultural systems. Crucially, we see this intervention as a tool to enable practitioners to move beyond viewing land as a static asset, and to instead begin analyzing the ways that land relations can be different. Centering questions of land tenure, access, and use also provides a generative pathway to map potential impacts of diverse agricultural interventions and how they will impact different groups in different ways. While it is all too common to see euphemistic or overly-sanguine promises about the potential impacts of a given intervention, focusing on land relations has the power to root these discussions in contextual realities and bring more sobering analysis to the realities of what development interventions can and cannot do.

2.4   Planting the Seeds of Change

The allegory of a gender tree helps to summarize these narratives, yet draws on fundamentally different paradigms and competing worldviews about what it takes to overcome poverty and inequality. The roots of the tree draw mainly on critical development theory and revolutionary change, while the branches of the tree draw on mainstream liberal paradigms focused on economic empowerment defined through asset accumulation. In writing this chapter and presenting the gender tree heuristic, our goal is not to subsume critical approaches within a liberal framework, but rather to emphasize the ways that these seemingly disparate paradigms do in fact intersect and influence each other. While practitioners often claim to be value-neutral, no intervention is free of theory, and while critical scholarship often denounces agricultural development practice, the seeds of these critiques have often punctured the world of practice in surprising ways. Critical and liberal approaches draw from radically opposed theories of change that call for divergent paths of action, from revolution to progressive reform. In writing this chapter, we do not seek to resolve this tension. Rather, we present a focus on land relations as a promising pathway to achieve better agricultural development results in the short term and lay the groundwork for more just and gender-equitable futures in the long term.

We hope the gender tree serves as a heuristic tool for doing the work of integrating liberal and critical bodies of work through the core question of land relations. In gender and agricultural research, critical approaches remind us to attend to the ways land is distributed, how access to usufruct and tenure rights are gendered, and often colonial, and how these patterns of land distribution and use shape the social realities in which agriculture research takes place and in which women farm. Effectively addressing the symptoms of gender inequality as liberal feminist approaches suggest, necessitates accounting for how women's access to land use and ownership is shaped by these historical and social processes. Our agricultural research will only ever yield partial explanations and remedies to gender inequality without a full accounting for gendered experiences of land relations.

We are hopeful that these conversations prompt reflection on the often-heard critique that critical theory is not concerned with impact, and that practice is not driven by theory. We see both of these coming through in the narratives we present: critical theories have intervened in agricultural development practice, and employing theoretical frameworks generates powerful impacts in agricultural development. We believe there is a need to be more explicit about the theories that inform practice. Yet we have also seen these critiques lose some of their critical edge as they are instrumentalized and depoliticized in practice. We acknowledge the potential that we are living through the very same phenomenon that we observed historically, as the critical research that generated participatory approaches and brought focus to women's empowerment were reformulated and robbed of their revolutionary potential in practice. Are intersectional and decolonial approaches in gender and agricultural development fated to be remembered with the same sense of instrumentalization and co-option? While this is certainly a risk, we draw inspiration from critical feminist scholarship that has repeatedly brought focus to the agency of women "beneficiaries" across the development landscape, and to the ways that these women rewrite their role in history on the land which sustains their lives.

Finally, at the methodological level, we see opportunities for the greater integration of critical and decolonial perspectives through the increasing use of feminist and community-engaged research methods across critical and practice-oriented research. In particular, we know that participatory methods are a valuable mode of engagement that create new spaces for reflection on social relations and the centering of community-based perspectives, such as ICRISAT's use of photovoice to evaluate nutrition impacts in Kenya (ICRISAT 2022). Participatory approaches readily center marginalized voices by design, and the lived experiences of intersectional identities bringing to light questions of land, power, and how and where people's lives are produced and reproduced (Brisolara et al. 2014). Recent feminist research in this field has called for the expansion of reflexive methodological approaches such as participatory photovoice (Photovoice, 2023) and photo elicitation as effective methods for understanding situated power relations within communities.

Although the demands of policy and practice tend to reduce diverse scholarship into "best practices" that address the visible manifestations of inequality, we wrote this chapter to call to the forefront the importance of *integrating* diverse perspectives in gender and agricultural development research by centering questions about land in this work. Who gets to produce on which types of land, and what types of land use are permitted in a particular socio-historical context? These are questions that gender and agricultural research must answer as fundamental background to all research addressing the technical manifestations of gender inequality. Today's global challenges necessitate the full integration of the possibilities for liberatory practice implied by decolonial and intersectional feminisms (Hooks 1991) and feminist solidarity across difference (Mohanty 2003). There is no universal, silver bullet approach to gender and agricultural development, but the future of the sector necessitates a way of thinking about situated interventions that take seriously questions of history, localized power relations and intersectional identities. These social dynamics play out in the history of land and on the contemporary use of land. We propose in this chapter a broader theory of change that emancipatory action is rooted in place, and to questions about land and social justice.

References

Abu-Lughod L (ed) (1998) Remaking women: feminism and modernity in the Middle East. Princeton University Press, Princeton

Adely F (2004) The mixed effects of schooling for high school girls in Jordan: the case of Tel Yahya. Comp Educ Rev 48(4):353–373. https://doi.org/10.1086/423361

Akram-Lodhi AH, Kay C (2010) Surveying the agrarian question (part 1): unearthing foundations, exploring diversity. J Peasant Stud 37(1):177–202. https://doi.org/10.1080/03066150903498838

Alarcón P (2022) Dependency revisited: Ecuador's (re)insertions into the international division of nature. Lat Am Perspect 49(2):207–226. https://doi.org/10.1177/0094582X211070831

Alkire S, Meinzen-Dick R, Peterman A, Quisumbing A, Seymour G, Vaz A (2013) The women's empowerment in agriculture index. World Dev 52:71–91. https://doi.org/10.1016/j.worlddev.2013.06.007

Altieri MA, Toledo VM (2011) The Agroecological revolution in Latin America: rescuing nature, ensuring food sovereignty and empowering peasants. J Peasant Stud 38(3):587–612. https://doi.org/10.1080/03066150.2011.582947

Anzaldúa G (1987) Borderlands/La Frontera: the new Mestiza. 25th anniversary, 4th edn. Aunt Lute Books, San Francisco

Badstue L, Elias M, Kommerell V, Petesch P, Prain G, Pyburn R, Umantseva A (2020) Making room for manoeuvre: addressing gender norms to strengthen the enabling environment for agricultural innovation. Dev Pract 30(4):541–547. https://doi.org/10.1080/09614524.2020.1757624

Benería L, Sen G (1981) Accumulation, reproduction, and "women's role in economic development": Boserup revisited. Signs J Women Cult Soc 7(2):279–298. https://doi.org/10.1086/493882

Bernstein H, Byres TJ (2001) From peasant studies to agrarian change. J Agrar Chang 1(1):1–56. https://doi.org/10.1111/1471-0366.00002

Bezner Kerr R, Hickey C, Lupafya E, Dakishoni L (2019) Repairing rifts or reproducing inequalities? Agroecology, food sovereignty, and gender justice in Malawi. J Peasant Stud 46(7):1499–1518. https://doi.org/10.1080/03066150.2018.1547897

Boserup E (1970) Woman's role in economic development. New York, St. Martin's Press

Brenner R (1976) Agrarian class structure and economic development in pre-industrial Europe. Past Present 70:30–75

Brisolara S, Seigart D, SenGupta S (2014) Feminist evaluation and research: theory and practice. Guilford Publications, New York

Carney J, Watts M (1990) Manufacturing dissent: work, gender and the politics of meaning in a peasant society. Africa J Int African Inst 60(2):207–241. https://doi.org/10.2307/1160333

Carr-Chellman DJ, Carr-Chellman A (2020) Integrating systems: the history of systems from von Bertalanffy to profound learning. TechTrends 64(5):704–709. https://doi.org/10.1007/s11528-020-00540-1

Collins PH, da Silva ECG, Ergun E, Furseth I, Bond KK, Martínez-Palacios J (2021) Intersectionality as critical social theory. Contemp Political Theory 20(3):690–725. https://doi.org/10.1057/s41296-021-00490-0

Cornwall A, Rivas A-M (2015) From "gender equality and women's empowerment" to global justice: reclaiming a transformative agenda for gender and development. Third World Q 36(2):396–415. https://doi.org/10.1080/01436597.2015.1013341

Crenshaw K (1990) Mapping the margins: intersectionality, identity politics, and violence against women of color. Stan L Rev 43:1241

DAWN Group (1986) Development crisis and alternative visions: third world women's perspectives. Canadian Woman Studies/les cahiers de la femme

Deere CD (2010) Household wealth and women's poverty: conceptual and methodological issues in assessing gender inequality in asset ownership. In: Chant S (ed) The international handbook of gender and poverty. Edward Elgar Publishing. https://doi.org/10.4337/9781849805162.00068

Dey J (1985) Women in African rice farming systems. In: Women in rice farming: proceedings of a conference on women in rice farming systems. International Rice Research Institute, Gower Publishers, Brookfield

Doss C (2013) Intrahousehold bargaining and resource allocation in developing countries. World Bank Res Obs 28(1):52–78. https://doi.org/10.1093/wbro/lkt001

Doss C, Meinzen-Dick R (2020) Land tenure security for women: a conceptual framework. Land Use Policy 99:105080

Doss C, Quisumbing A (2021) Gender, household behavior, and rural development. Part three: context for agricultural development, chapter 15. In: Otsuka K, Fan S (eds) Agricultural development: new perspectives in a changing world. International Food Policy Research Institute (IFPRI), Washington, DC, pp 503–528. https://doi.org/10.2499/9780896293830_15

Doss C, Summerfield G, Tsikata D (2014) Land, gender, and food security. Fem Econ 20(1):1–23

Doss C, Meinzen-Dick R, Quisumbing A, Theis S (2018) Women in agriculture: four myths. Glob Food Sec 16:69–74. https://doi.org/10.1016/j.gfs.2017.10.001

FAO (2011) The state of food and agriculture. Food and Agriculture Organization, Rome

FAO (2023) The status of women in agrifood systems. Food and Agriculture Organization, Rome

Farhall K, Rickards L (2021) The "gender agenda" in agriculture for development and its (lack of) alignment with feminist scholarship. Front Sustain Food Syst 5:573424. https://doi.org/10.3389/fsufs.2021.573424

Faxon HO (2020) Securing meaningful life: women's work and land rights in rural Myanmar. J Rural Stud 76:76–84

Feldstein HS (1998) An inventory of gender-related research and training in the Consultative Group on International Agricultural Research (CGIAR) Centers. CGIAR

Feldstein HS, Poats S (1989) Working together: gender analysis in agriculture V.1 case studies. Kumarian Press, West Hartford

Ferguson J (1990) The anti-politics machine: "development" depoliticization, and bureaucratic power in Lesotho. Cambridge University Press, Cambridge

Ferguson AE (1994) Gendered science: a critique of agricultural development. Am Anthropol 96(3):540–552. https://doi.org/10.1525/aa.1994.96.3.02a00060

Fukuda-Parr S (2016) From the millennium development goals to the sustainable development goals: shifts in purpose, concept, and politics of global goal setting for development. Gend Dev 24(1):43–52. https://doi.org/10.1080/13552074.2016.1145895

Geisler G (1993) Silences speak louder than claims: gender, household, and agricultural development in Southern Africa. World Dev 21(12):1965–1980. https://doi.org/10.1016/0305-750X(93)90069-L

Goodman D, Watts M (eds) (1997) Globalising food: agrarian questions and global restructuring. Routledge, London

Hajjar R, Ayana AN, Rutt R, Hinde O, Liao C, Keene S, Bandiaky-Badji S, Agrawal A (2020) Capital, labor, and gender: the consequences of large-scale land transactions on household labor allocation. J Peasant Stud 47(3):566–588

Hall D (2013) Primitive accumulation, accumulation by dispossession and the global land grab. Third World Q 34(9):1582–1604. https://doi.org/10.1080/01436597.2013.843854

Hart G (2001) Development critiques in the 1990s: culs de sac and promising paths. Prog Hum Geogr 25(4):649–658. https://doi.org/10.1191/030913201682689002

Hooks B (1991) Theory as liberatory practice. Yale J Law Fem 4(1) https://openyls.law.yale.edu/handle/20.500.13051/7151

ICRISAT (2022) Using photovoice to assess impacts of behavior change nutrition activities using in Kenya – ICRISAT. https://www.icrisat.org/using-photovoice-to-assess-impacts-of-behavior-change-nutrition-activities-in-kenya/

Jiggins J (1986) Gender-related impacts and the work of the international agricultural research centers. CGIARhttps://hdl.handle.net/10947/582

Johnson NL, Kovarik C, Meinzen-Dick R, Njuki J, Quisumbing A (2016) Gender, assets, and agricultural development: lessons from eight projects. World Dev 83:295–311. https://doi.org/10.1016/j.worlddev.2016.01.009

JP GTA (2022) Joint programme on gender transformative approaches for food security and nutrition. https://www.fao.org/joint-programme-gender-transformative-approaches/en

Kabeer N (1991) Gender, development, and training: raising awareness in the planning process. Dev Pract 1(3):185–195

Kabeer N (1999) Resources, agency, achievements: reflections on the measurement of women's empowerment. Dev Chang 30(3):435–464

Kandiyoti D (1998) Rethinking bargaining with the patriarchy. In: Jackson C, Pearson R (eds) Feminist visions of development: gender analysis & policy. Routledge, New York

Kawarazuka N, Prain G (2019) Gendered processes of agricultural innovation in the northern uplands of Vietnam. Int J Gend Entrep 11(3):210–226. https://doi.org/10.1108/IJGE-04-2019-0087

Kingiri AN (2013) A review of innovation systems framework as a tool for gendering agricultural innovations: exploring gender learning and system empowerment. J Agric Educ Ext 19(5):521–541. https://doi.org/10.1080/1389224X.2013.817346

Leder S, Sachs CE (2019) Intersectionality at the gender–agriculture nexus: relational life histories and additive sex-disaggregated indices. In: *Gender, agriculture and agrarian transformations*. Routledge, London, pp 75–92

Li TM (2007) The will to improve: governmentality, development, and the practice of politics. Duke University Press, Durham

Li TM (2014) Land's end: capitalist relations on an indigenous frontier. Duke University Press, Durham

Majeedullah A, Wied K, Mills E (2016) Gender, sexuality and the sustainable development goals: a meta-analysis of mechanisms of exclusion and avenues for inclusive development. IDS Evidence Report 206

Malapit H, Quisumbing A, Meinzen-Dick R, Seymour G, Martinez EM, Heckert J, Rubin D, Vaz A, Yount KM (2019) Development of the project-level women's empowerment in agriculture index (pro-WEAI). World Dev 122:675–692. https://doi.org/10.1016/j.worlddev.2019.06.018

Manji AS (2006) The politics of land reform in Africa: from communal tenure to free markets. Zed, London

Meinzen-Dick RS, Quisumbing AR, Doss CR, Theis S (2017) Women's land rights as a pathway to poverty reduction: a framework and review of available evidence. IFPRI discussion paper 1663, Washington DC. http://ebrary.ifpri.org/cdm/ref/collection/p15738coll2/id/131359

Mintz SW (1986) Sweetness and power: the place of sugar in modern history. Penguin, London

Mohanty CT (1984) Under Western eyes: feminist scholarship and colonial discourses. Boundary 2:333–358

Mohanty CT (2003) "Under Western eyes" revisited: feminist solidarity through anticapitalist struggles. Signs J Women Cult Soc 28(2):499–535

Moser C (1993) Gender planning and development: theory, practice and training (1st ed.). Routledge. https://doi.org/10.4324/9780203411940

Moser C, Moser A (2005) Gender mainstreaming since Beijing: a review of success and limitations in international institutions. Gend Dev 13(2):11–22

Mukhopadhyay M, Wong F (eds) (2007) Revisiting gender training: the making and remaking of gender knowledge. A global sourcebook. KIT (Royal Tropical Institute)/Oxfam, Amsterdam/Oxford

Okali C (2012) Gender analysis: engaging with rural development and agricultural policy processes. January. https://opendocs.ids.ac.uk/opendocs/handle/20.500.12413/2318

Overholt C, Anderson M, Cloud K, Austin J (1985) Gender roles in development projects. Kumarian Press, West Hartford

Oxfam (2014) Gender action learning system. Oxfam Novib. https://www.oxfamnovib.nl/Redactie/Downloads/English/publications/150115_Practical%20guide%20GALS%20summary%20Phase%201-2%20lr.pdf

Photovoice (2023) Photovoice website. https://photovoice.org/

Puskur R, Mudege NN, Njuguna-Mungai E, Nchanji E, Vernooy R, Galiè A, Najjar D (2021) Moving beyond reaching women in seed systems development. In: Pyburn R, van Eerdewijk A (eds) Advancing gender equality through agricultural and environmental research: past, present, and future. International Food Policy Research Institute, Washington, DC, pp 113–145. https://doi.org/10.2499/9780896293915_03

Pyburn R, van Eerdewijk A (2021) Advancing gender equality through agricultural and environmental research: past, present, and future: synopsis. International Food Policy Research Institute, Washington, DC. https://doi.org/10.2499/9780896294202

Quisumbing AR, Meinzen-Dick R, Raney TL, Croppenstedt A, Behrman JA, Peterman A (eds) (2014) Gender in agriculture: closing the knowledge gap. Springer, Dordrecht. https://doi.org/10.1007/978-94-017-8616-4

Rao A, Kelleher D (2005) Is there life after gender mainstreaming? Gend Dev 13(2):57–69

Sachs CE (1996) Gendered fields: rural women, agriculture, and environment, Rural studies series. Westview Press, Boulder

Sachs CE (ed) (2019) Gender, agriculture and agrarian transformations: changing relations in Africa, Latin America and Asia, Earthscan food and agriculture series. Routledge, London

Sachs CE, Jensen L, Castellanos P, Sexsmith K (eds) (2020) Routledge handbook of gender and agriculture. Routledge, London. https://doi.org/10.4324/9780429199752

Saito KA, Mekonnen H, Spurling D (1994) Raising the productivity of women farmers in sub-Saharan Africa. World Bank discussion papers, Africa Technical Department Series 230, Washington DC

Sandoval C (1991) US third world feminism: the theory and method of oppositional consciousness in the postmodern world. Genders 10:1–24

Sardenberg CMB (2008) Liberal vs. liberating empowerment: a Latin American feminist perspective on conceptualising women's empowerment. IDS Bull 39(6):18–27

Sen G, Grown C (1987) Development, crises and alternative visions: third world women's perspectives. NYU Press, New York

Sent EM, van Staveren I (2019) A feminist review of behavioral economic research on gender differences. Fem Econ 25(2):1–35

Slavchevska V, Doss CR, de la O Campos AP, Brunelli C (2021) Beyond ownership: women's and men's land rights in sub-Saharan Africa. Oxf Dev Stud 49(1):2–22

Stromquist NP (1999) The theoretical and practical bases for empowerment. Women, education, and empowerment: pathways towards autonomy, pp 13–22

Sud N, Sánchez-Ancochea D (2022) Southern discomfort: interrogating the category of the Global South. Dev Chang 53(6):1123–1150. https://doi.org/10.1111/dech.12742

Tarjem IA, Westengen OT, Wisborg P, Glaab K (2023) "Whose demand?" The co-construction of markets, demand and gender in development-oriented crop breeding. Agric Hum Values 40:83–100. https://doi.org/10.1007/s10460-022-10337-y

Tavenner K, Crane TA (2022) Hitting the target and missing the point? On the risks of measuring women's empowerment in agricultural development. Agric Hum Values 39(3):849–857. https://doi.org/10.1007/s10460-021-10290-2

Tavenner K, Crane TA, Bullock R, Galiè A (2022) Intersectionality in gender and agriculture: toward an applied research design. Gend Technol Dev 26(3):385–403. https://doi.org/10.1080/09718524.2022.2140383

Taylor I (2016) Dependency redux: why Africa is not rising. Rev Afr Polit Econ 43(147):8–25. https://doi.org/10.1080/03056244.2015.1084911

Theis S, Meinzen-Dick R (2016) Reach, benefit, or empower: clarifying gender strategies of development projects. https://a4nh.cgiar.org/2016/11/29/reach-benefit-or-empower-clarifying-gender-strategies-of-development-projects/

Tufan HA, Mangheni MN, Boonabaana B, Asiimwe E, Jenkins D, Garner E (2021) GREAT expectations: building a model for applied gender training for crop improvement. J Gend Agri Food Secur (Agri-Gender) 6(2):1–18. https://doi.org/10.19268/JGAFS.622021.1

van der Burg M (2020) Gender integration in international agricultural research for development. Routledge handbook of gender and agriculture. Routledge, Abingdon, pp 69–84

Walinsky L, Ladejinsky W (1993) Agrarian reform as unfinished business. Oxford University Press, New York

Whitehead A (2000) Continuities and discontinuities in political deconstructions of the working man in rural sub-Saharan Africa: the 'lazy man' in African agriculture. Eur J Dev Res 12(2):23–52

Wolford W (2010) This land is ours now: social mobilization and the meanings of land in Brazil. Duke University Press, Durham, NC

World Bank, Food and Agriculture Organization, and International Fund for Agricultural Development (2009) Gender in agriculture sourcebook. The World Bank, Washington, DC

Zakaria R (2021) Against white feminism: notes on disruption. Norton & Company, New York

Chapter 3
Visualizing the Gendering of Agricultural Mechanization in the Global South: A Review of the Underlying Drivers

Moses Kansanga ⓘ and Dinko Hannan Dinko ⓘ

Abstract While mechanization has emerged as a key pathway for improving agriculture in the Global South, the materialization of its perceived benefits has been hindered by gender inequalities in technology use. Deeply connected to this gendering of technology are debates about the drivers of women's low access to and control of mechanized agricultural technologies. Drawing on the gender mechanization literature and based on insights from mechanization research in sub-Saharan Africa, we explore the multi-scalar factors that explain the gendering of mechanized technologies in the Global South and provide policy recommendations. Overall, our analysis contributes to the gender-mechanization literature by demonstrating that the masculinization of agricultural technologies is not just a product of the widely discussed role of structural factors in local agrarian spaces, but an element of the very design and engineering of mechanized technologies that attribute masculine traits to them, and the consistent deployment of gender (in)sensitive agricultural programs by governments and development partners. Given that masculinization of technology is endemic in agriculture and transcends mechanization, we identify governments and development partners as viable intermediaries who are well positioned to channel feedback on the needs of women to the agricultural machinery industry upstream, while ensuring a gender-sensitive deployment of mechanization services downstream.

M. Kansanga (✉)
George Washington University, Washington, DC, USA
e-mail: mkansanga@email.gwu.edu

D. H. Dinko
Mount Holyoke College, South Hadley, MA, USA
e-mail: dhanaandinko@mtholyoke.edu

3.1 Introduction

Debates about addressing global hunger have always referred to the pivotal role of technology in improving productivity. In the Global South in particular, mechanized technology deployment is seen as a pathway to closing yield gaps and addressing drudgery in smallholder agriculture (Cele 2021; Daum and Birner 2020; Paudel et al. 2019). The Green Revolution in Asia and Latin America and the ongoing new Green Revolution for Africa are testaments to the reliance on technology and innovation. Through these programs, mechanized technologies such as tractors, seed sowers, sprayers, and combine harvesters have been widely deployed by governments in partnership with international development organizations as part of national agricultural development programs (Dinko 2017). Scholarship on the impacts of these technologies has however revealed diverse social concerns including gender inequalities in technology control and access, which continue to work against the materialization of the perceived benefits of these programs (Carney and Watts 1991; Gengenbach et al. 2017; Kansanga et al. 2019). These emerging gender-related concerns are crucial given the pivotal role women play in smallholder agriculture in the Global South.

Although there is a consensus in the technology adoption literature that women farmers are less likely to adopt agricultural technologies than male farmers (Paudel et al. 2020; Polar et al. 2017; Theis et al. 2018a), this discrepancy has sparked new questions about what makes agricultural technologies gender-biased. While some scholars theorize this gendered adoption of technology to be an outcome of women's differential technology adoption decision making (e.g. Owusu et al. 2018), others explain this imbalance in technology access and use to be an outcome of women's sociocultural positioning in agrarian communities and the associated farm roles, resources access opportunities and abilities attached to these positions (Badstue et al. 2020; Carney and Watts 1991; Van Eerdewijk and Danielsen 2015; Kansanga et al. 2019; Mohammed et al. 2023; Paudel et al. 2020). Others point to the potential broader environment and political economy within which agricultural technologies are imagined, designed, and deployed (Aryal et al. 2019; Quaye et al. 2021; Vemireddy and Choudhary 2021). The physical design of technologies, including their size and structure, may also espouse some male attributes which can feed into local spaces where they are considered the domains of men and not women. Some scholars have called for gender-sensitive technology models such as smaller horsepower and two-wheel tractors which can be easily operated by women (Baudron et al. 2015; Ngoma et al. 2023). This raises the argument that technologies themselves can be simultaneously empowering and disempowering, depending on their design.

Drawing on the gender-mechanization literature, this chapter takes a holistic approach in progressively contextualizing the gendering of agricultural technology access and control by working outward from the local to the global. Locally, the aim is to draw on empirical case studies on gender and technology adoption in smallholder farming communities in the Global South to visualize the gendered politics of technology adoption. Attention will be paid to how women's cultural, agrarian roles and resource rights mediate their access to agricultural technologies. The role of national policy and of machinery manufacturers will also be explored, along with the part played by global policy actors like the World Bank.

3.2 Methods

We started with a rapid review of the literature. Unlike a systematic review, a rapid review provides more timely evidence to support decision-making. A rapid review allows for a compelling overview of the vast and quickly evolving literature on agricultural mechanization and gender dynamics, to guide policy appraisal and scholarly discourse (Davis et al. 2009; Pham et al. 2014; Kerr et al. 2022; Khangura et al. 2012; Sharpe et al. 2017).

In spite of its advantages, rapid reviews have several limitations. First, rapid reviews are less systematic and thus likely to miss some publications which may otherwise have been included in a systematic review. Secondly, due to the relatively limited search scope, rapid reviews are exposed to publication bias (Moons et al. 2021). For example, in this chapter, our search was limited to papers published in English, inadvertently excluding relevant studies written in other languages. The narrow scope of the review necessitates a cautious approach when interpreting the findings (Harker and Kleijnen 2012).

Our discussion of gender and mechanization in the developing world avoids specific cross-country comparisons. We synthesize the literature on the Global South as a geographic realm. The rapid review was done in three stages. Stage (1) identifying the research question: *What empirical evidence exists on the intersections of gender and mechanization in the Global South?* Stage (2) conducting systematic searches in five electronic databases: Ebscohost, Google Scholar, Agora, Web of Science, and GreenFILE. A combination of Boolean Operators (e.g., AND, OR, NOT, AND NOT and "") as conjunctive and or exclusionary terms were used to narrow or expand search results. For example ["gender" "women" "mechanization" AND Africa OR Asia]. We limited the search to the last three decades (1990–2022). Prior dates are unlikely to reflect current scholarship on gender and mechanization in the Global South given recent, rapid changes (Fig. 3.1). The papers were downloaded and organized according to geographic location.

Overall, 350 papers were identified to be relevant to the research question after initial sorting. We then screened the abstracts of the 350 to reduce the number to a manageable size of 117. Papers were included if they explicitly and empirically addressed gender and mechanization in agriculture either as a case study or a review. Papers were excluded if they made only commentaries without empirical or traceable data. After the second round of screening, 66 papers passed the test of relevance and were read in full and charted.

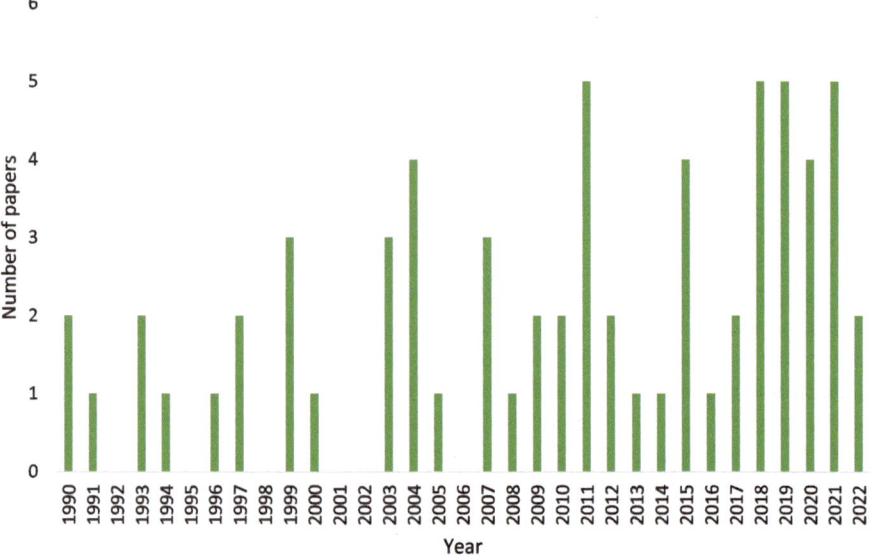

Fig. 3.1 Summary of papers reviewed by year of publication

3.3 Results

Three drivers of gendered technology adoption emerged from this review: 1) local level gender dynamics (100%, *n = 66*) 2) national level dynamics of technology deployment (68%, n = 45) and 3) broader political economy of agricultural policy design, targeting, and implementation (57%, n = 38). While each paper made a unique contribution to the discourse on gender and mechanization, most of them identified more than one of the three drivers, so there is some double counting. All papers cited are part of the n = 66. As Fig. 3.2 from the 66 papers shows, these three factors interact to reinforce gendered technology adoption among smallholder farmers in the global south. The processes of scaling (upscaling and downscaling) in political economy generally denotes the interaction of causal mechanisms or actors operating from the local to the national or the international (in the case of upscaling) or in reverse order from the international to the local.

Through downscaling, national, and global level agricultural policies privilege export-oriented crops (13.6% of papers *n = 9*) as entry points of technology introduction in farming. Export-oriented farming is often dominated by men who tend to have secure land tenure due to patrilineal land customs and norms (e.g., Amanor and Iddrisu 2022; Carney and Watts 1991; Saliou et al. 2020). This male dominance is then taken as the standard and default in designing and manufacturing farm machines, tools, and implements, creating an implicit perception of male suitableness in the adoption of mechanization (Andersson et al. 2022; Carney and Watts 1991; Devkota et al. 2022; Paudel et al. 2020). These perceptions are then exploited

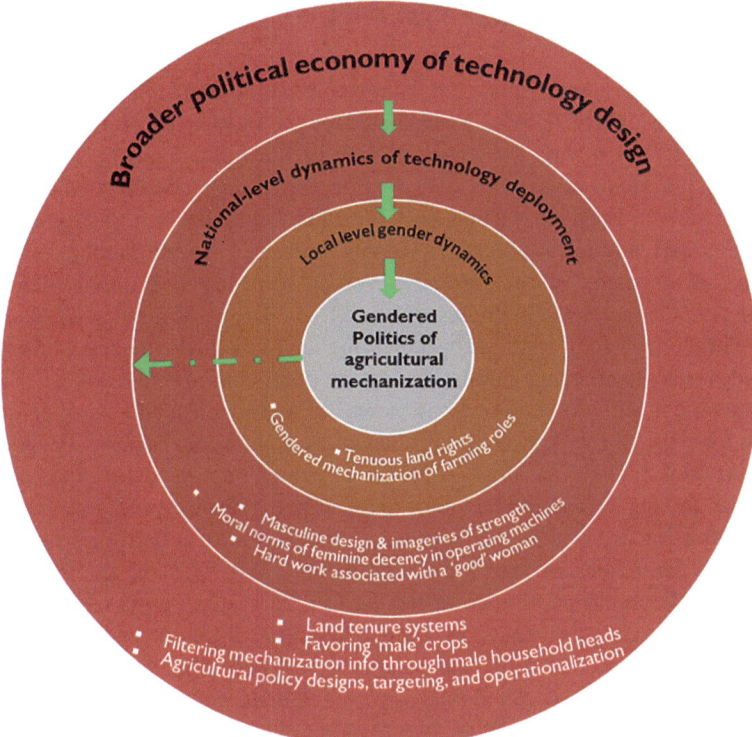

Fig. 3.2 Interrelating factors that shape the gendered adoption of mechanization among smallholder farmers in the Global South

to reinforce local cultural and gender norms and practices on women's place and work in society.

Through upscaling, local gender dynamics also feed into and shape the design and manufacturing of labor-saving technologies. For instance, the predominant local decision-making norms inform which farming tasks should be mechanized. There is longstanding evidence that men tend to make decisions on what mechanized tools to buy, as well as decisions on when to use such tools and machines (Carney and Watts 1991; Paudel et al. 2020; Polar et al. 2017; Theis et al. 2018a; Vemireddy and Choudhary 2021; Wiig 2013). This decision-making privilege is often taken as the industry standard in manufacturing implements and thus farm machine manufacturers target men in advertising new technologies. The male privilege in agricultural technology adoption further feeds into the broader pollical economy of policy making, targeting and implementation.

We discuss how the three factors identified in the literature review shape the gendering of mechanization in the Global South. We contextualize these themes using examples from the literature.

3.3.1 Local Level Gender Dynamics

Sociocultural norms and intrahousehold power dynamics tend to privilege the dominance of male decisions on what mechanized technology can be adopted, how much is used, when, where, and on whose portion of the farm. Theis et al. (2018a) in examining the gender dynamics of irrigation technology adoption among smallholder farmers in Ethiopia, Ghana, and Tanzania found that historically engrained land rights and cultural notions of a woman's place in the household constrained adoption of mechanization technology. Women were less likely than men to use mechanized irrigation technologies (Theis et al. 2018a). Kansanga et al. (2019) observed that male farm roles such as plowing tend to be mechanized while female farm roles such as planting, and harvesting are often entirely manual.

Njuki et al. (2022) have shown how norms of feminine decency have proscribed women from using treadle water pump irrigation in Tanzania, because it exposes the outline of their thighs. Kansanga (2017) has documented how social expectations of hard work and endurance and the quest to avoid being labeled lazy drives women to endure labor-intensive farm activities to fit the portrait of a hardworking woman, even as the farm roles of men have been mechanized. Theis et al. (2018b) have shown how the taboo on women operating rice harvesters limits opportunities to mechanize women's activities in Bangladesh. In discussing the intersections between gender, mechanization, and intrahousehold time allocation on farms, Daum et al. (2020), argue that so-called men's crops and male-dominated activities tend to be prioritized when a household decides to adopt mechanization. This is largely due to the gendered power imbalances in patriarchal societies.

Van Eerdewijk and Danielsen (2015) argue that women have little power in decisions over mechanized technology adoption. Thus, investments in farm machinery are more likely to save men's labor than to alleviate women's suffering (Doss 2002). Badstue et al. (2020) highlight how social norms of submission and reproductive roles give men authority over what women can and are allowed to do and how that profoundly affects the adoption of new mechanized technologies in Kenya and Ethiopia. These findings were echoed in India (Mehta et al. 2018), Nepal (Paudel et al. 2019, 2020), Zambia (Adu-Baffour et al. 2019), and Peru (Wiig 2013) where sociocultural norms that privilege male decision-making at the household level limit women from adopting mechanized technologies.

3.3.2 National-Level Dynamics of Mechanized Technology Policy Formulation and Deployment

Aside from sociocultural and masculine exclusionary designs of farm machines, the broader national level policies and how they are implemented have implications for gendered agricultural mechanization. Agricultural policy design, targeting, and operationalization can result in a gendered uptake of mechanization that inadvertently disadvantage women. For example, Theis et al. (2018b) and Doss (2002) have shown that when technological inputs are tied to farm size and land ownership, women are less likely to benefit.

In the Global South, most land tenure systems put ownership rights and management decisions in the hands of men. Although women may have use rights, they are often secondary and predicated on men granting such rights. Women tend to have smaller and marginal land than men. Tenure rights affect the gendered adoption of technology in two ways. First, tenuous tenure discourages women from investing in long-term labor-saving technologies which may require an initial capital cost. With no certainty of access for the next farming season, short-term adoption of labor-saving technology with no guarantee of continued use could be prohibitive. Second, returns to investments are lower on women's smaller farms. Hence, access to mechanization is tied to farm size and land ownership, limiting opportunities for women's adoption. Yet, government policies have largely been passive towards these concerns with flagship agricultural development policies either excluding women from technology support programs through inclusion criteria founded on land ownership or deploying technologies to perform farm tasks not usually done by women. As highlighted by Farnworth and Colverson (2015), governments and development partners have complicated and deepened this gendering in their deployment of technology services by habitually defining a "farmer" with male attributes such as land ownership, which automatically disqualifies most women.

Government mechanization programs in the Global South have often targeted commercial and export crops (Vemireddy and Choudhary 2021). For instance, in Ghana, Amanor and Iddrisu (2022) demonstrate how mechanization programs that provided threshers and plows prioritized cotton production. Omulo et al. (2022) have also shown how mechanization tends to promote medium-scale and market-oriented farmers who are often men. There is also growing evidence that mechanization programs that promote specific crops often lead to male invasion of hitherto "women crops." For example, Tsusaka et al. (2016) have shown how mechanization and commercialization of groundnut in Zambia and Malawi led to men taking over groundnut cultivation. Adu-Baffour et al. (2019) have similarly linked mechanization initiatives in Zambia to men displacing women farmers.

The gendered politics of agricultural mechanization also reflect the technology adoption process. Theis et al. (2018b) identify three phases in farm technology adoption: awareness, trying out, and continued adoption. Women are systematically disadvantaged in all three phases. For instance, information and communication bottlenecks such as not owning radio sets or limited literacy can limit women's access to information on new labor-saving tools and machinery. Even when women gain access to information, their ability to try out new technologies may be hindered by a lack ownership and control over land and money necessary to adopt these technologies. When the first two hurdles are cleared, uncertain land tenure (Kansanga et al. 2019), affordability (Afridi et al. 2020; Carney and Watts 1991), cultural appropriateness (Aryal et al. 2019; Farnworth et al. 2022; Theis et al. 2018b) and return-to-investments (Daum and Birner 2017, 2020) could limit technology adoption.

3.3.3   The Broader Political Economy of Technology Design

The physical design and structure of farm machineries such as tractors, tillers, and harvesters have been criticized for being masculine and not female-friendly (Theis et al. 2018a). Tractors are often associated with masculine images of strength and power, albeit that may be changing gradually due to inventions like power steering which makes tractors easier to operate (Cele 2021). In the Global South where manual tractors are still dominant, operating tractors require physical strength which tends to be associated with men. Thus, some tractors require certain muscle strength to operate them, excluding women. For instance, Cele (2021) has highlighted how perceived masculine strength and physical power hinder women's participation as tractor operators in Ghana and limit "womechanisation". In the Global North, scholars have stressed how notions of femininity and masculinity shape how farm women use and talk about tractors (Laszlo Ambjörnsson 2021; Newsome 2021; Pini 2005).

3.4   Discussion and Next Steps

While there is increasing emphasis on using machines to improve smallholder agriculture in the Global South, in practice most mechanization programs have favored men farmers. Gender inequalities in agricultural technology access and control are usually theorized as resulting from micro-level structural factors that disadvantage women. While local gender norms do influence the gendering of technology access in the Global South, this paper highlights the crucial role of two other drivers that are rarely discussed: (1) male-biased technology deployment by governments and international development partners and (2) farm equipment is built with masculine traits by manufacturers.

We argue that agricultural technologies by their design reinforce existing inequalities; technologies come with masculine traits that discourage already marginalized women from operating them. Technology production is cast in a broader political economy of invention that does not consider the everyday needs of women. This gender (in)sensitive technology design is largely a result of an enduring disconnect between machine manufacturers and farmers, governments and development partners (Andersson et al. 2022; Devkota et al. 2022; Fischer et al. 2021). This is not to suggest that local dynamics do not contribute to the gendering of agricultural technology access and control. As demonstrated by this review, local agricultural norms, including the gendered division of labor and women's culturally ascribed positions in households limit their access to mechanized technologies. Governments habitually set inclusion criteria for subsidized mechanized services based on male-dominant structures such as land ownership. This is exemplified by the case study in Ghana where available mechanized technologies including tractors and herbicides only substitute the culturally ascribed farm roles of men (e.g., Amanor and Iddrisu 2022; Dinko and Nyantakyi-Frimpong 2023; Kansanga et al. 2019; Cele 2021).

Thus, while local dynamics are central to the gendering of technology access, we argue that the broader political economy of mechanization and related-gender exclusionary technology design are also a crucial source of these gender inequalities. Machinery from factories already come with masculine attributes, for example, their large size and the physical strength needed to operate them. This masculine structure then feeds into local social construction of the use of these technologies where, for example, it is seen as odd for a woman to operate a tractor but "normal" for a man (Devkota et al. 2022; van Eerdewijk and Danielsen 2015). Governments and international development partners are key intermediaries between machinery designers and manufacturers, with a better opportunity than farmers to communicate feedback on gender inequality concerns, yet technologies are consistently deployed within environments of inequality.

Given the findings in this paper we can suggest some next steps to make mechanization more inclusive. First, although there are multiple entry points for addressing the gendering of agricultural technology in the Global South, governments and development partners remain a strategic link for undoing this engrained bias both upstream and downstream. Governments and development partners, with their better links to manufacturers, should convey gender -based concerns with farm machinery companies and dealers. This way, mechanized technologies reaching farming communities will reflect everyday gender realities and not just masculine attributes.

Second, governments and development partners must examine existing and future mechanized technology support programs to farmers to ensure they are comprehensive and respond to the full breadth and depth of labor needs in farming communities.

Third, active gender-transformative education should accompany technology support programs to create spaces for women to renegotiate cultural expectations around women's ability to operate machines. Such programs should also address the longstanding access limitations for women that are created by cultural norms. While women are not passive recipients of gender-insensitive mechanization, gender-transformative programing is necessary to provide platforms for women's collective organizing and renegotiation of gender biases in agricultural technology access and control.

Finally, there is also a place for gender scholars working on agriculture to continually think about these issues and provide the conceptual framework for future research guiding action on these themes to ensure that newer sources of gender inequality do not emerge.

References

Adu-Baffour F, Daum T, Birner R (2019) Can small farms benefit from big companies' initiatives to promote mechanization in Africa? A case study from Zambia. Food Policy 84:133–145. https://doi.org/10.1016/j.foodpol.2019.03.007

Afridi F, Bishnu M, Mahajan K (2020) Gendering technological change: evidence from agricultural mechanization. www.iza.org

Amanor KS, Iddrisu A (2022) Old tractors, new policies and induced technological transformation: agricultural mechanisation, class formation, and market liberalisation in Ghana. J Peasant Stud 49(1):158–178. https://doi.org/10.1080/03066150.2020.1867539

Andersson K, Pettersson K, Lodin JB (2022) Window dressing inequalities and constructing women farmers as problematic—gender in Rwanda's agriculture policy. Agric Hum Values 39(4):1245–1261. https://doi.org/10.1007/s10460-022-10314-5

Aryal JP, Rahut DB, Maharjan S, Erenstein O (2019) Understanding factors associated with agricultural mechanization: a Bangladesh case. World Dev Perspect 13:1–9. https://doi.org/10.1016/j.wdp.2019.02.002

Badstue L, van Eerdewijk A, Danielsen K, Hailemariam M, Mukewa E (2020) How local gender norms and intra-household dynamics shape women's demand for laborsaving technologies: insights from maize-based livelihoods in Ethiopia and Kenya. Gend Technol Dev 24(3):341–361. https://doi.org/10.1080/09718524.2020.1830339

Baudron F, Sims B, Justice S, Kahan DG, Rose R, Mkomwa S, Kaumbutho P, Sariah J, Nazare R, Moges G, Gérard B (2015) Re-examining appropriate mechanization in Eastern and Southern Africa: two-wheel tractors, conservation agriculture, and private sector involvement. Food Secur 7:889–904

Carney J, Watts M (1991) Disciplining women? Rice, mechanization, and the evolution of Mandinka gender relations in Senegambia. J Women Cult Soc 16(4):651–681

Cele L (2021) Empowerment and agricultural mechanisation: perceptions and experiences of women tractor operators in Ghana. Dev Pract 31(8):988–1001. https://doi.org/10.1080/09614524.2021.1937551

Daum T, Birner R (2017) The neglected governance challenges of agricultural mechanisation in Africa—insights from Ghana. Food Secur 9(5):959–979. https://doi.org/10.1007/s12571-017-0716-9

Daum T, Birner R (2020) Agricultural mechanization in Africa: myths, realities and an emerging research agenda. Glob Food Sec 26:100393. https://doi.org/10.1016/j.gfs.2020.100393

Daum T, Adegbola YP, Kamau G, Kergna AO, Daudu C, Zossou RC, Crinot GF, Houssou P, Mose L, Ndirpaya Y, Wahab AA, Kirui O, Oluwole FA (2020) Impacts of agricultural mechanization: evidence from four African countries. Hohenheim Working Papers on Social and Institutional Change in Agricultural Development

Davis K, Drey N, Gould D (2009) What are scoping studies? A review of the nursing literature. Int J Nurs Stud 46(10):1386–1400. https://doi.org/10.1016/j.ijnurstu.2009.02.010

Devkota R, Pant LP, Hambly Odame H, Rai Paudyal B, Bronson K (2022) Rethinking gender mainstreaming in agricultural innovation policy in Nepal: a critical gender analysis. Agric Hum Values 39(4):1373–1390. https://doi.org/10.1007/s10460-022-10326-1

Dinko DH (2017) Theory and practice: Changing faces of rural development policy in Ghana from 1957-2007. African Journal of Agricultural Economics and Rural Development 5(3):539–546. https://www.airitilibrary.com/Article/Detail/P20160728001-201703-201711240019-201711240019-539-546

Dinko DH, Nyantakyi-Frimpong H (2023) Uneven geographies of the embodied effects of water insecurity among women irrigators in Northern Ghana. Ann Am Assoc Geogr 113:2417–2434. https://doi.org/10.1080/24694452.2023.2231528

Doss CR (2002) Men's crops? Women's crops? The gender patterns of cropping in Ghana. World Dev 30(11):1987–2000. https://doi.org/10.1016/S0305-750X(02)00109-2

Farnworth CR, Colverson KE (2015) Building a gender-transformative extension and advisory facilitation system in sub-Saharan Africa. J Gend Agric Food Secur 1(1):20–39

Farnworth CR, Bharati P, Krishna VV, Roeven L, Badstue L (2022) Caste-gender intersectionalities in wheat-growing communities in Madhya Pradesh, India. Gend Technol Dev 26(1):28–57. https://doi.org/10.1080/09718524.2022.2034096

Fischer G, Kotu B, Mutungi C (2021) Sustainable and equitable agricultural mechanization? A gendered perspective on maize shelling. Renew Agric Food Syst 36(4):396–404. https://doi.org/10.1017/S1742170521000016

Gengenbach H, Schurman RA, BassettTJ MWA, Moseley WG (2017) Limits of the New Green Revolution for Africa: Reconceptualising gendered agricultural value chains. Geogr J 184(2):208–214

Harker J, Kleijnen J (2012) What is a rapid review? A methodological exploration of rapid reviews in health technology assessments. Int J Evid Based Healthc 10(4):397–410. https://doi.org/10.1111/j.1744-1609.2012.00290.x

Kansanga MM (2017) Who you know and when you plough? Social capital and agricultural mechanization under the new green revolution in Ghana. Int J Agric Sustain 15(6):708–723. https://doi.org/10.1080/14735903.2017.1399515

Kansanga MM, Antabe R, Sano Y, Mason-Renton S, Luginaah I (2019) A feminist political ecology of agricultural mechanization and evolving gendered on-farm labor dynamics in northern Ghana. Gend Technol Dev 23(3):207–233. https://doi.org/10.1080/09718524.2019.1687799

Kerr RB, Liebert J, Kansanga M, Kpienbaareh D (2022) Human and social values in agroecology: a review. Elem Sci Anth 10(1):00090. https://doi.org/10.1525/elementa.2021.00090

Khangura S, Konnyu K, Cushman R, Grimshaw J, Moher D (2012) Evidence summaries: the evolution of a rapid review approach. Syst Rev 1(1):1–19. https://doi.org/10.1186/2046-4053-1-10

Laszlo Ambjörnsson E (2021) Performing female masculinities and negotiating femininities: challenging gender hegemonies in Swedish forestry through women's networks. Gend Place Cult 28(11):1584–1605. https://doi.org/10.1080/0966369X.2020.1825215

Mehta CR, Gite LP, Khadatkar A (2018) Women empowerment through agricultural mechanization in India. Curr Sci 114(9):1934–1940

Mohammed K, Batung E, Saaka SA, Kansanga MM, Luginaah I (2023) Determinants of mechanized technology adoption in smallholder agriculture: implications for agricultural policy. Land Use Policy 129:106666. https://doi.org/10.1016/j.landusepol.2023.106666

Moons P, Goossens E, Thompson DR (2021) Rapid reviews: the pros and cons of an accelerated review process. Eur J Cardiovasc Nurs 20(5):515–519. https://doi.org/10.1093/eurjcn/zvab041

Newsome L (2021) Disrupted gender roles in Australian agriculture: first generation female farmers' construction of farming identity. Agric Hum Values 38(3):803–814. https://doi.org/10.1007/s10460-021-10192-3

Ngoma H, Marenya P, Tufa A, Alene A, Chipindu L, Matin MA, Thierfelder C, Chikoye D (2023) Smallholder farmers' willingness to pay for two-wheel tractor-based mechanisation services in Zambia and Zimbabwe. J Int Dev 35:2107–2128. https://doi.org/10.1002/jid.3767

Njuki J, Eissler S, Malapit H, Meinzen-Dick R, Bryan E, Quisumbing A (2022) A review of evidence on gender equality, women's empowerment, and food systems. Glob Food Sec 33:100622. https://doi.org/10.1016/j.gfs.2022.100622

Omulo G, Daum T, Köller K, Birner R (2022) Are emerging farmers the missing link for mechanised conservation agriculture? Viewpoints from Zambia. Dev Pract 32(3):411–417. https://doi.org/10.1080/09614524.2022.2036702

Owusu AB, Yankson PW, Frimpong S (2018) Smallholder farmers' knowledge of mobile telephone use: gender perspectives and implications for agricultural market development. Prog Dev Stud 18(1):36–51. https://doi.org/10.1177/1464993417735389

Paudel GP, Bahadur KCD, Rahut DB, Justice SE, McDonald AJ (2019) Scale-appropriate mechanization impacts on productivity among smallholders: evidence from rice systems in the midhills of Nepal. Land Use Policy 85:104–113. https://doi.org/10.1016/j.landusepol.2019.03.030

Paudel GP, Gartaula H, Rahut DB, Craufurd P (2020) Gender differentiated small-scale farm mechanization in Nepal hills: an application of exogenous switching treatment regression. Technol Soc 61:101250. https://doi.org/10.1016/j.techsoc.2020.101250

Pham MT, Rajić A, Greig JD, Sargeant JM, Papadopoulos A, Mcewen SA (2014) A scoping review of scoping reviews: advancing the approach and enhancing the consistency. Res Synth Methods 5(4):371–385. https://doi.org/10.1002/jrsm.1123

Pini B (2005) The third sex: women leaders in Australian agriculture. Work Organizat 12(1):73–88

Polar V, Babini C, Velasco C, Flores P, Fonseca C (2017) Technology is not gender neutral: factors that influence the potential adoption of agricultural technology by men and women (La Paz- Bolivia). International Potato Center, Lima. chrome-extension://efaidnbmnnnibpcajpcglc lefindmkaj/https://mail.wocan.org/sites/default/files/Technology%20is%20not%20gender%20 neutral%20ENG.pdf

Quaye W, Onumah JA, Boimah M, Mohammed A (2021) Gender dimension of technology adoption: the case of technologies transferred in Ghana. Dev Pract 32(4):434–447. https://doi.org/1 0.1080/09614524.2021.2000588

Saliou IO, Zannou A, Aoudji AKN, Honlonkou AN (2020) Drivers of mechanization in cotton production in Benin. West Africa Agric 10(11):549. https://doi.org/10.3390/agriculture10110549

Sharpe EE, Karasouli E, Meyer C (2017) Examining factors of engagement with digital interventions for weight management: rapid review. JMIR Res Protoc 6(10):e205. https://doi.org/10.2196/resprot.6059

Theis S, Lefore N, Meinzen-Dick R, Bryan E (2018a) What happens after technology adoption? Gendered aspects of small-scale irrigation technologies in Ethiopia, Ghana, and Tanzania. Agric Hum Values 35(3):671–684. https://doi.org/10.1007/s10460-018-9862-8

Theis S, Sultana N, Krupnik TJ (2018b) Overcoming gender gaps in rural mechanization: lessons from reaper-harvester service provision in Bangladesh. No. 9. International Food Policy Research Institute (IFPRI)

Tsusaka TW, Orr A, Msere HW, Homann-Keetui S, Maimisa P, Twanje GH, Botha R (2016) Do commercialization and mechanization of a "women's crop" disempower women farmers? Evidence from Zambia and Malawi. Agricultural and Applied Economics Association (formerly the American Agricultural Economics Association), pp 1–26

Van Eerdewijk A, Danielsen K (2015) Gender matters in farm power. KIT, CIMMYT and CGIAR, Amsterdam

Vemireddy V, Choudhary A (2021) A systematic review of labor-saving technologies: implications for women in agriculture. Glob Food Sec 29:100541. https://doi.org/10.1016/j.gfs.2021.100541

Wiig H (2013) Joint titling in rural Peru: impact on women's participation in household decision-making. World Dev 52:104–119. https://doi.org/10.1016/j.worlddev.2013.06.005

Chapter 4
What Happened to Gender-Intentional Participatory Research in Agriculture?

Jacqueline A. Ashby

Abstract The chapter analyzes the evolution of participatory research in agriculture and how gender and gender analysis have been used within it. In the late twentieth century, farmer participation in agricultural research was a strategy for social justice, to share power with the rural poor and organize them as a client base. However, feminist research in the 1990s noted that participatory research often reinforced gender norms in agriculture. In the 1980s participatory research was incorporated into the international agricultural research system. In the 1990s gender mainstreaming was adopted in the international centers. However, these science bureaucracies soon divorced gender and participatory research from any political engagement. Research managers saw participatory research, and gender analysis, as ways to design better farm technology, and ease its dissemination. By 2010, participatory plant breeding was widely adopted. In the early 2010s the system again became interested in gender equality. There is now a need for participatory research to become fully gender-intentional, to empower women and farming communities. The large-scale transformation of agriculture calls for a fundamental change in how research engages with its client base, among the rural poor.

4.1 Introduction

Participatory approaches to improving agriculture that involve and empower women are considered essential for progress in ameliorating food insecurity, low productivity, poor diets, and undernutrition (FAO 2023). Globally, among the 300 million people currently estimated at high risk of acute food insecurity, a gender gap persists: 32% of women are moderately or severely food-insecure compared with 28% of men (FAO et al. 2022). Food insecurity is more prevalent in countries with high gender inequality (FAO, 2018; Harris-Fry et al. 2020). Farms managed by women are 24% less productive that those of the same size managed by men (FAO 2023).

J. A. Ashby (✉)
International Development Consulting, Portland, OR, USA
e-mail: jacqueline.ashby@cantab.net

© The Author(s) 2025   51
J. Njuki et al. (eds.), *Gender, Power and Politics in Agriculture*,
https://doi.org/10.1007/978-3-031-60986-2_4

According to FAO (2023), removing gender inequalities in agricultural productivity and wages would increase global GDP by almost US\$ 1 trillion and reduce the number of food insecure people by about 45 million. In sum, improving efficiency is a widely promoted rationale for participatory approaches that empower women and for using gender-intentional participatory research (PR) in agriculture. However, the alignment of participation and empowerment with the requirements of research institutions is criticized for diluting these concepts. Critics argue that the way participation and empowerment are operationalized, measured and implemented to fit with positivist science does not improve empowerment or gender equality (Tavenner and Crane 2022). Gender-intentional participatory approaches have been co-opted and depoliticized and so require "re-politicizing" with a renewed commitment to transformative change(Clisby and Enderstein 2017; De Jong and Kimm 2017; Acosta et al. 2019; Jackson and Pearson 2005; Cornwall 2003; Cornwall and Rivas 2015; Njuki et al. 2016).

Gender-intentional participatory research in agriculture refers to the use of participatory research that is informed by analysis of constraints that operate differently for man and women, including gender norms and relationships and other relevant gender inequalities, with the intention of addressing them. This chapter examines what happened to participatory research in agriculture, and gender within it, to analyze the shortcomings that give rise to the call for re-politicization. The setting for this analysis is the multi-institution international agricultural research system consisting of the research centers of the Consultative Group on International Agricultural Research (CGIAR), the national agricultural research and extension systems (NARES) in over 100 countries, regional research networks, universities in all countries, non-government organizations (NGOs) and farmer organizations (FOs) and the Global Forum on Agriculture (GFAR). In 2022 their research partnerships encompassed 95% of the global area growing the 21 crops, forages, livestock and fish species that CGIAR works on (CGIAR Initiative on Market Intelligence 2024). The international agricultural research system is comprised primarily of public-sector research organizations (although these partner widely with the private sector) and their donors, which are national governments and international foundations.

The chapter examines the role in this story of two aspects of the depoliticization of gender-intentional participatory research in agriculture: the divorce from action and the failure to build an organized client base. De-politicization is the result of the appropriation and reinterpretation of key concepts of participation, gender and empowerment in ways that diluted their original purpose and blunted their impact. This process was not unique to agricultural research; what happened there was a microcosm of the larger processes of the incorporation of participation, gender and empowerment by influential development agencies and donors, the recent interest in citizen science and espousal of transformative approaches to development (De Jong and Kimm 2017; Burns et al. 2021; Batliwala 2002; Lengwiler 2008). The various stages of gender mainstreaming and of participatory research implementation were not always or even often, aligned with each other. An important outcome has been the divorce of participation and gender analysis from action research and from transformative approaches to empowerment. The next section reviews the institutional contact and history, followed by a discussion of the two main features of depoliticization singled out for analysis, and what these may mean for future "re-politicization."

4.2 Background: Institutional Context and History

Participation in research initially gained prominence in rural development circles in the late twentieth century as a strategy for the poor to gain social justice (Leal 2010; Freire 1970). From this perspective, participation in research is fundamentally a renegotiation of who has agency and whose knowledge is legitimate. In knowledge-generation, participation involves power-sharing among experts and lay people in setting a research agenda, prioritizing a problem, identifying a solution, and taking action to implement it. Participation in research was formulated as one of several steps in a political process by activists (Gramsci 1971; Freire 1970; Fals-Borda 1987) who argued that different kinds of knowledge and ways of knowing are central to differences in power and help to propagate social injustice. In their view, participatory research should be paired with political action that empowers participants to achieve a more just society. From this political perspective, informing action to change power relations is the overriding goal of participatory research and the rationale for the term "participatory action research" (Chambers and Thrupp 1994; Chambers 1997).

Feminist research has an explicitly political agenda: gender relations are deeply embedded in the agency of individuals and groups and therefore, influence the distribution and exercise of power in knowledge creation. Thus, gender relations necessarily color the nature, extent, and outcomes of participation in research. However, this perspective was not widely applied in development or agricultural research until the late 1990s when the focus of institutional gender mainstreaming shifted from integrating women into development (WID) towards one concerned with gender relations (Jackson and Pearson 2005; Miller and Razavi 2011). Although feminist scholars demonstrated how gender shapes knowledge and concomitant power over resources (Rocheleau et al. 1996), feminist theory in the 1990s did not integrate its concern for social justice with participatory action research until much later (Lykes et al. 2007). Instead, feminist researchers pointed out how a wide array of participatory approaches reinforced gender norms that restrict rural women's participation and empowerment (Cornwall 2003).

Participatory research in agriculture originated in participatory action research, but acquired a different set of objectives in scientific research organizations. In agriculture, participatory research refers to an eclectic set of approaches, methods, and tools for enabling farmer-scientist interaction that began to attract scientific attention in the 1980s and now are in use, in one form or another, in a variety of public- and private-sector programs all over the globe. The number and diversity of organizations involved is vast. Participatory research has been used by organizations like the Food and Agriculture Organization (FAO), the World Bank, the US Department of Agriculture (USDA), the European Union, and in more than 60 national programs and innumerable NGOs in projects to promote technology adoption in Europe, North and South America, Asia, including China, and throughout Africa (Tribaldos et al. 2020; Ceccarelli and Grando 2020; Waters-Bayer et al. 2015). Application of PR has involved thousands of resource-poor producers, men and women, in low-income countries who rely on traditional farming methods as well as hundreds of large-scale farmers in high income countries (Tribaldos et al. 2020). As of 2019, participatory approaches in plant breeding had been used in 69 countries (10 developed and 59 developing) with 47 crops in international research centers, national programs, universities and NGOs (Ceccarelli and Grando 2019). Participatory research methods have been used to include poor men and women in the development of new plant varieties, innovative seed systems, sustainable intensification farming practices, soil conservation, water management, animal husbandry, food safety and crop protection and small machinery (Johnson et al. 2003; Ceccarelli and Grando 2020; Tribaldos et al. 2020; Ashby 2009; Scoones et al. 2009; Waters-Bayer et al. 2015; Humphries et al. 2012).

Although feminist research and participatory action research share the goal of changing prevailing power relations, the two streams of thought were not fully integrated in the 1990s when participatory research and gender mainstreaming took root in separate institutional niches in the international agricultural research system. For pioneers in participatory research who found common cause in the 1987 workshop leading to the publication of the book Farmer First (Chambers and Thrupp 1994), inequality between men and women was less of a concern than reversing top-down relationships that put the rural poor, both men and women, at the bottom of hierarchies of power. They noted that the objective of giving farmers a voice in development required attention to gender (Chambers and Thrupp 1994; Chambers 1997). Nonetheless, a decade later, the 1997 World Congress of Participatory Action Research did not devote space to feminist perspectives or representation to women (Fals-Borda, 1998).

Participatory research and gender were incorporated into the international agricultural research system against this backdrop, and it is useful to have an overview of the timeline. This chapter covers the period from the 1980s to 2020. In the 1980s concepts of participation and gender were novel–even radical—in the system. Throughout this period they were promoted from different institutional entry points in different research centers. The Women and Rice Farming Systems Project (1985) and the Intra-Household and FSRE Case Studies Project (1984–1994) were the first CGIAR initiatives concerned with women in agriculture. Elsewhere in 1982 a small participatory research group formed around social scientists at the International Potato Center (CIP), who published the "farmer-back-to-farmer" model (Rhoades and Booth 1982). This model contained the seeds of the participatory technology evaluation approaches developed a few years later at the International Center for Tropical Agriculture (CIAT) that formed the backbone of support for participatory plant breeding by the CGIAR Systemwide Program on Participatory Research and Gender Analysis (PRGA) funded in 1996, a time when global interest in participation and empowerment as development strategies were at a height (Narayan and Petesch 1997). For its first ten years, the PRGA prioritized participatory plant breeding and participatory natural research management.

By the mid-2000s participatory research had gained institutional recognition and was cited in the 2008 World Development Report: Agriculture for Development (World Bank 2007). By the end of the 2000s, participatory plant breeding continued to spread, reaching 60 NARES by the next decade (Ceccarelli and Grando 2020) while in the CGIAR, participatory evaluation of varieties was adopted as a tool that most breeding programs should have in their portfolio (Walker 2006). Inside and outside of agriculture, participatory approaches became a standard feature of global climate change and natural resource management.

Gender research in the international agricultural research system had a different trajectory. It divided into different initiatives with different leadership and funding, one concerned with institutional gender mainstreaming, another focused on policy-oriented social research and a third concerned with integrating gender analysis into biological, technical research to improve the gender-relevance of technology development (Poats 1990). Institutional gender mainstreaming took off when donor responses to the 1986 CGIAR Impact Study (Jiggins 1986) called for further, system-level attention to gender. Donors ensured that from 1991–1995 the system's central Secretariat hosted a Gender Program that supported mainstreaming in workplace and research (CGIAR 1995). In 1996, the Secretariat closed the Gender Program: its gender research component was included in the PRGA program, while policy-oriented research on gender developed at another CGIAR Center. This consolidated a divergence in the system between policy-oriented research vs applied research concerned with systemic integration of gender into technology development. Policy-oriented research produced a rich body of influential social research publications, but it did not engage with the technical agenda of the system's biological scientists, and applied technical gender research lagged in the face of institutional push-back until 2012 (van der Burg 2019).

After 2006, PRGA prioritized mainstreaming gender-sensitive participatory research through training and mentoring for NARES, until it closed in 2010. A PRGA diagnostic study concluded that interest in participatory approaches was increasing but gender issues remained " largely ignored" (Alvarez et al. 2010). Nonetheless, during the 2000s national gender policies propelled investment in gender mainstreaming by NARES and regional agricultural research networks. Donor interest in gender in international agricultural research reignited with the publication of FAO (2011) and World Bank (2012) reports on gender inequality: from 2012–2018 the CGIAR System Management Office obtained a mandate for oversight of all research plans of work and budget, including gender research, which enabled investment in applying gender to technical research in all CGIAR programs. In 2019, responsibility for gender in research moved out of the system's central administration, this time to a Platform, GENDER, that spans the entire CGIAR, with a mandate to synthesize and amplify gender in research.

In contrast, many NGOs integrated gender-intentional participatory research with their overall empowerment objectives. They formed grassroots farmer groups to work with scientists and extensionists, supported by farmer cooperatives, networks and federations. Some were political movements such as Campesino a Campesino which currently involves two million families worldwide, and MASIPAG, a farmer-led network of over 500 grassroots groups, 60 NGOs and 15 scientific organizations working for sustainable management of diversity through farmers' control of genetic resources, agricultural production and associated knowledge. Other NGOs promoted grassroots innovation through participatory research, such as the multi-stakeholder platforms in 19 countries affiliated with Participatory Research and Innovation for Sustainable Livelihoods (PROLINNOVA) (Waters-Bayer et al. 2009), the Local Agricultural Research Committees (CIAL) and similar experimenting farmer groups (Ashby et al. 2000; Humphries et al. 2012; Ashby 2009) in or integrated farmer-led research in local development, e.g., World Neighbors in Bolivia, Mali and Honduras. Another approach was associated with farmer field school programs, some of which included farmer-led experimentation in addition to their usual discovery-learning demonstration plots (Van den Berg et al. 2020). Farmer-led experimentation strengthened local capacity to innovate but found little opportunity for institutional partnerships with formal research organizations (Waters-Bayer et al. 2015).

Collective action by rural people was central for NGOs that retained the goal of informing action through learning and knowledge generation. Pretty et al. (2020) identified a wide range of social movements, networks and federations among rural people involved in collaborative efforts to support sustainable agriculture and social equity. Their analysis calculated that the number of intentionally-formed collaborative groups in 122 sustainable agriculture and environmental initiatives in 55 countries had grown from half a million in 2000 to 8.54 million in 2020, with 170–255 million group members, farming an area of about 300 million hectares. The groups' improvements in agricultural productivity were achieved by farmers and land managers working with scientists and extensionists (Pretty et al. 2020).

Farmer-led research in grassroots collaborative groups acquired many different forms and technical objectives, such as the farmer field schools that sought policy action, experimenting farmer groups and women's self-help groups that prioritized agricultural experimentation (Van den Berg et al. 2020; Brummett and Jamu 2011; Humphries et al. 2012; Ashby et al. 2000). Their common ground was that the roles of scientists, extensionists and other technical advisors were transformed, as were the status and agency of participating farmers (Pretty et al. 2020; Almekinders et al. 2006; Rosset and Martínez-Torres 2014). In the grassroots institutional setting, gender-intentional participatory research subscribed to holistic models of agricultural science such as agroecology or permaculture and emphasized the validity of indigenous forms of discovery as distinct from the scientific method (Altieri 1989; Richards 1985; Altieri and Toledo 2011; Altieri and Nicholls 2017). Critically, in grassroots organizations this work was aligned with a political agenda seeking food sovereignty and social justice for poor farmers, such as access to land, water and sustainable technologies. Grassroot organizations pursued disruptive and transformative innovation in the social structures underpinning social and gender inequality (Bebbington et al. 2005; Waters-Bayer et al. 2015).

The rest of this chapter looks at two major aspects of this history.

4.3   Divorce from Action Research

Gender-intentional participatory research in agriculture evolved as adaptations by international research organizations, which define themselves as science bureaucracies. Gender-intentional participatory research influenced how these bureaucracies responded to innovation. Science bureaucracies are organized as a hierarchy of specialized roles and functions, filled by individuals who are not elected but appointed. Work is done according to an established system of rules, regulations and policies that is intended to be apolitical and impersonal, in the sense that the work performed does not require direct contact with those being served, the clients (Louis and Maertens, 2021). Power to make decisions, authority and accountability in a bureaucracy are defined by a hierarchical chain of command and control (Blau and Meyer 1971).

At the same time, bureaucracies frequently have a culture that prizes flexibility in enforcing regulations, tacit knowledge, and reliance on disseminating ideas and information through informal channels. In research organizations this culture can foster innovation, but has been shown to enable the persistence of gender-blind attitudes and practices that hinder gender mainstreaming (Roth and Sonnert 2011). The renegotiation of norms and re-invention of ideas in the bureaucracies' informal culture, described by Cleaver (2017) as institutional bricolage, can support innovation, but can also be a form of cooptation. In science, normative practices can be concealed within apparently objective research methods. Accepted routines for making decisions and setting priorities also function as what Jasanoff terms "repositories of power" (Jasanoff 2002). In this chapter science bureaucracy refers to public-sector organizations whose primary purpose is scientific research, as distinct from other types of organization, mostly NGOs or FOs that engage in gender-intentional, farmer-led participatory research.

Agricultural science bureaucracies' preference for biology and related disciplines fostered a reinterpretation of social concepts of participation, empowerment, and gender equality (Ashby 2009). The scientific method with its focus on explanatory reduction and its standards of empirical proof were an epistemological environment in which action researchers with empowerment as a goal had little legitimacy. Their response was to amplify the efficiency rationale for participatory research and gender analysis. One justification was to lower research costs, so increasingly tight operating budgets could be stretched when farmers were responsible for all operations in the test fields (Walker 2006). Participatory research is still widely understood as a means of improving the reliability of recommendations originating from on-station research (Laajaj et al. 2020). Another justification was the idea systematized and popularized by farmer first pioneers, that low adoption of post-green revolution technology was the result of bypassing farmers' indigenous knowledge. Early farmer first work was heavily focused on showing that new technologies were more likely to satisfy the preferences of poor people when scientists paid attention to farmer's advice. The need to overcome hierarchical power relations within science bureaucracies that reinforced skepticism about farmer knowledge and innovation meant that the early work of farmer first pioneers was focused on demonstrating the importance of discrepancies between scientists' and farmers' views of technology, so that differences between men and women's adoption choices took second place (Chambers and Thrupp 1994; Scoones and Thompson 2009).

In a similar process, the justification for concern with gender from an efficiency perspective distanced gender in agricultural research from feminist political goals and social justice for farm women. The efficiency rationale enabled practitioners to make a non-threatening case with an instrumental perspective that gender experts used to promote their case to senior management and donors (e.g., Jiggins 1986). Recognition of gender-differentiated farming practices had roots in farming systems research in the 1970s to understand the lack of fit between recommended technologies and location-specific conditions. Preoccupation with the yield-gap between on-station and on-farm performance of post-green revolution technologies provoked interest in achieving a gender balance in farming systems research that explains the early entry point gained by the Women and Rice Farming Systems program (WIRF 1991). Attention to gender was promoted in relation to research representativity, access by researchers to women informants' opinions and advice, and how gender could influence farmers' opinions (Chambers and Thrupp 1994; Feldstein 2000). Practitioners' adoption of the efficiency rationale in science bureaucracies reflected, and was reinforced by, wide use of the same argument in international development circles (Cornwall and Brock 2005; FAO 2011; World Bank 2012). The first impact assessment of CGIAR concluded that the rationale for attention to gender was to realize the full potential impact of research on output, productivity and welfare (Jiggins 1986; Herdt and Anderson 2019).

Reliance on the efficiency rationale led rapidly in the mid-1990s to divorce from action research with political objectives. For gender and participation in agriculture, as in the wider development agency and donor context, this meant arguments for empowerment and transformative change dropped out of sight (Moser and Moser 2005; Moser 2005). What persisted was a mainstreaming approach that emphasized gender balance in the workplace and gender-sensitivity training for researchers, while policy-oriented research on gender developed independently of the CGIAR's GENDER program. This was not at this time, successful in defining a clear application to technology development for gender-sensitivity or gender research, so that even sympathetic technical scientists perceived it as irrelevant to their work.

For participatory research, the divorce from action became institutionalized in the 1990s with the distinction between researcher-led and farmer-led participatory research. Participatory research in agriculture relies on combining local or indigenous knowledge of farmers with normal science conducted by agricultural scientists for the co-production of new knowledge. This approach, termed "researcher-led participatory research," includes some kinds of citizen science: it is initiated by scientists who open up spaces for consultation and collaboration with farmers and other stakeholders who have less power in the research decisions than scientists

(Lilja and Bellon 2008; Kimura and Kinchy 2016). A different approach is farmer-led, similar to a more egalitarian citizen-scientist alliance, in which scientists act as resource-persons and have less power than farmers who control the agenda (Lilja and Ashby 1999; Johnson et al. 2003; Bidwell 2009).

Evidence from a variety of participatory research processes shows that who defines the agenda and problem, and who initiates the participation is a key difference among approaches, with enduring consequences for the authority and credibility of the results (Mosse 2003). For example, after visiting and analyzing numerous participatory plant breeding (PPB) projects that had been running for about 10 years, Almekinders et al. (2006) concluded that there were two approaches based on the type of actor taking the lead: one approach led by plant breeding programs prioritized technical outcomes, i.e., developing improved varieties. The second approach, led mainly by NGOs and FOs, prioritized empowerment of farmers to strengthen their self-reliance, maintain genetic diversity in farmer's fields and safeguard their ownership over their genetic resources.

By subscribing wholeheartedly to the efficiency rationale, practitioners endorsed a researcher-led approach and deprived themselves of an effective challenge to the dominant "diffusion of technology" paradigm. This paradigm, popularized in the 1950s, puts technical experts (researchers or extensionists) and farmers in a doctor-patient relationship: the experts have a duty of care to protect farmers from risk by prescribing technologies and practices validated by on-station research. When adoption rates lag, diffusion of technology defines the problem as the farmers' behavior, attitudes and skills or the farmers' environment, not the supply of technology nor the researchers' responsibility.

In summary, once divorced from action research with its political empowerment objectives, the application of participation, empowerment and gender equality experienced mission-drift. Farmer-led approaches that opened up significant agency in research decisions about technology design for farm men and women, had difficulty in gaining an institutional footing compared with researcher-led approaches such as mother-baby trials that rely on the doctor-patient relationship (Ceccarelli and Grando 2019). Integration of gender struggled to find an application in technical research beyond simple sex-disaggregation and involvement of women respondents in participatory fieldwork (van der Burg 2019).

4.4   Divorce from an Organized Client Base

Throughout the international agricultural research system, a widely accepted hierarchy of research disciplines placed biological science on a superior footing in relation to social science. Biological scientists accepted social scientists' proposition of learning from men and women farmers to improve research efficiency. But in the 1990s, more radical implications of gender-intentional participation, were inconceivable, such as direct accountability to men and women producers, processors and consumers for delivering adoptable technology. In the absence of accountability to farmers, biological scientists could argue that influencing gender equality and empowerment was outside their scope of influence and not their responsibility.

Lack of accountability to farmers meant that from the start, the international agricultural research system had built-in a disconnect between its supply of technology and demand from the rural poor. Technology design was decided in laboratories and research stations. Technology testing on farms seldom invited opinions from farmers. Women were rarely identified as farmers. Historically weak or absent linkages between supply and demand for agricultural technologies meant that participatory research practitioners experienced a divided identity, they had both a vested interest in promoting the technology supplied by research, and also were responsible for empowering participating farmers with agency in research (Resurrección and Elmhirst 2020; Pohl et al. 2010; Ashby 2009; Dewulf et al. 2005; Rist et al. 2006; Ashby 2009). Critical empirical studies of participatory action research showed that a facilitator's vested interest in an outcome motivated participants to distort information to fit researchers' expectations (Cooke and Kothari 2001; Kothari 2001).

The gender-sensitivity of participatory methods depended vitally on the user's awareness and understanding of gender (Cornwall 2003; Mosse 1994). Acceptance of the divorce from action research contributed to role conflict that occurs when an individual has to follow contradictory norms and "rules of the game." Critics argued that authentic co-production of knowledge is not compatible with positivist research standards and researchers consequently experienced divided identity (Gaventa and Cornwall 2008; Ravetz 2001, p. 391; Fals-Borda 1987). For many practitioners, co-existence with supply-side technology development, the diffusion of technology paradigm, and lack of accountability to farmers contradicted the principles of participatory research. Role conflict at the individual level contributed to disunity and fragmentation of perspectives at the group level: individuals attempted to resolve contradictions each in their own way by tweaking methods that emphasized to varying degrees, researcher-led or farmer-led principles and gender-aware or gender-transformative objectives.

Lack of cohesion in perspectives made cooperation more challenging for social scientists engaged in participatory research or gender analysis in agriculture. This was partly the result of their focus on innovating with methods and tools. Embracing the efficiency rationale and abandoning action research with a political purpose led to a focus on developing methods and tools instead of seeking structural change. By the 1990s there was a clear divide: tools and methods were acceptable research products to donors and science managers, whereas building an organized client base among farm people was not.

The implication for gender within participatory research was an early focus on compiling gender facilitation guides, tools and toolboxes. The notion was lost that participation requires empowerment as a foundation for gendered knowledge production (Tavenner and Crane 2022; Sumberg et al. 2013).

By the early 2000s, the widespread and growing perception among donors as well as researchers that concerns for participation, gender and empowerment belonged in NGOs and not in scientific research, contributed to the fragmentation of approaches. From early days, lack of cohesion around an agreed set of standards for what constituted scientific application of participatory research and gender analysis to technology design and delivery diluted intellectual leadership for gender-intentional participatory research. Lack of an agreed set of consistent standards enabled mission-drift that was hard to detect or to correct. This enabled managers' indiscriminate use of "participatory" and "gender-sensitive" in research proposals to donors, who contributed to the lack of cohesion by rewarding the gratuitous use of rhetoric with funding. Dilution of leadership and fragmentation of effort made replication difficult and undermined scaling up of a coherent approach for working with biological scientists. Impactful gender research that scaled out, such as development of WEIA, was independent conceptually and financially of technology development in the system.

Adaptation to bureaucratic norms and culture produced an approach to participation, gender equality and empowerment divorced from action research, lacking accountability to client farmers, without agreed standards or methods, and hampered by dilution of leadership and institutional fragmentation. All of the above contributed to the inability of researchers outside the NGO sector to institutionalize a client base among poor farmers with effective channels for expressing demand. Since the 1980s, hundreds of thousands of short-term, transitory farmer meetings and groups were convened solely for research purposes. These congregations are different from the 8.5 million groups intentionally-formed for long-term collective action, learning or land care documented by Pretty et al. (2020). Although impact studies conducted in the early 2000s concluded that participatory research could provide a basis for building collective farmer organization at national levels (Eade and Friis-Hansen 2008; Humphries et al. 2012) donors and biological scientists demarcated building farmer organizations as a job for NGOs, not for researchers.

The importance of building the client base of PR to deliver benefits to poor men and women farmers is illustrated by the example of sorghum PPB in Burkina Faso from 2002–2020 (Vom Brocke et al. 2020). A thorough impact study concluded that the program had increased famers income and food security as well as strengthened skills, social capital and organizational strength in farmer seed Unions (Christinck et al. 2014; Vom Brocke et al. 2020; Rattunde et al. 2021). Starting in the 1990s, sorghum breeders embraced paradigm change, discovering that women farmers were reluctant to grow the existing improved varieties which were unsuited to low soil fertility, especially common on plots of poor women. Breeders not only went back to the drawing board to make new crosses, they also began PR that strengthened the professionalization of farmer seed unions, especially encouraging women's membership, ensuring effective delivery of PPB varieties to the client base and creating a grassroots farmer lobby that succeeded in influencing national seed policy. This outcome was enabled by the breeders' immersion in a stakeholder-driven process with relationships built from farm to policy levels throughout the value chain at national and regional scales. Their commitment to organizing farmers was supported by a donor invested in building the grassroots organization required for empowering participation (Christinck et al. 2014).

However, most frontline researchers were discouraged from building the client base for PR, because few science bureaucracies and their donors saw this as an appropriate outcome for scientific research. Even though politically influential farmer organizations and NGOs gained a foothold in policy-making bodies, such as GFAR, this could not remedy the lack of investment in building a grassroots client base with a strong voice in setting research priorities. Research providers throughout the international research system continued to sidestep accountability to farmers. The absence of accountability is convenient for research providers, but antithetical to the idea that participation should be empowering, giving voice as well as the authority to sanction (Sen 1988).

Weak linkages to demand from an organized client-base of farmers allowed science bureaucracies to defuse the power reversals inherent in giving farmers agency and power through participation in research decisions. This was reflected in the slow recognition of empowerment as an outcome for which research could accept responsibility. As early as the 1970s, social inclusion and gender equity were important in farming systems research, based on the argument that poor farmers in marginal, low-resource farming environments were not benefitting from green revolution and later innovations (Sumberg and Okali 1997). In the late 1980s, gradients of empowerment were distinguished to differentiate the extent to which farmers could participate in research decisions. Method and tool development were tied to conceptual distinctions between empowering versus non-empowering participation (Pretty 1995; Biggs 1989).

The inclusion of women was integrated into these approaches and methods drawing on the then current women-in-development perspective (Lilja and Ashby 1999). Even though empowerment of farm women as decisionmakers was put forward as a research impact by gender researchers, this kind of impact did not gain a foothold in evaluation, and empowering farmers was deemed a by-product of research (Walker 2006; Paris 2008). Although gender equity and empowerment objectives were widely advocated in development circles, especially after publication of the World Development Report on Gender Equality (World Bank 2012), in agricultural science bureaucracies empowerment of women did not gain credibility as a desirable research outcome until the 2020s. Even then, the main preoccupation was inclusion that covered all the social categories (e.g., age, sex, poverty, ethnicity). Inclusive participation with a gender perspective was expanded to include entire value chains. The emphasis on inclusion caused many biological scientists to view concerns with gender and with participation as the same thing, and to perceive both as ways of conducting market research (Tarjem et al. 2022).

The goal of empowering farmers fueled efforts by NGOs to organize farmer-led research groups. These groups commonly set their own research agenda, had some control over their own research funds and were linked in innovation networks, such as the Honey Bee network in India, PROLINNOVA networks in Ethiopia, the CIAL national network in Honduras, Colombia and elsewhere in Latin America (Abrol and Gupta 2014; Waters-Bayer et al. 2015; Humphries et al. 2012; Ashby et al. 2000; Almekinders et al. 2006; Rosario 2009; Moreira and Mulvany, 2009; Almekinders and Elings 2001). Gender inequalities tended to emerge as an issue in farmer-led research when the facilitators included women in farmer-led research groups and made sure women's opinions were heard and responded to. Over time, many such farmer-led networks took up the empowerment of women as an explicit objective and developed leadership by women that had important benefits (Waters-Bayer et al. 2015; Humphries et al. 2012; Almekinders et al. 2006). There were valuable complementarities between farmer-led research and networks of self-help groups, including the thousands of women's self-help groups formed by micro-finance (de Boef et al. 2021). In Asia, sub-Saharan Africa and Latin America, women's self-help groups took part in varietal evaluations, seed production and delivery (Isaacs et al. 2024; de Boef et al. 2021; Christinck et al. 2014; Rattunde et al. 2021; Benitez Fernández et al. 2023). Empowerment, especially of women, gained significance in the research process, but protagonists of farmer-led research were unable to convince science bureaucracies that empowering poor rural producers was a precondition for authentic feedback to research.

Empowerment had a different trajectory where farmers were better educated, wealthier and better organized. For example, in Europe, North America, the United Kingdom, Brazil and India where large-scale programs were established with funding for farmer-led, participatory research and support for networks promoting farmer-to-farmer innovation. In the USA, the USDA supported farmer-led experimentation nationwide that was widely influential in developing conservation agriculture, with the Sustainable Agricultural Research and Education program (SARE) started in 1988 (Gliessman and Rosemeyer 2009). In the European Union, national and regional networks supporting farmer-led research for several decades, have had both influential political lobbies and a role in innovation for sustainable farming. The agroecology networks supported by the Latin American Consortium on Agroecology and Sustainable Development (CLADES), an NGO, had viable links to grassroots organizations within active political movements.

4.5 Conclusions

The requirements and culture of positivist science in formal research organizations led to the appropriation and reinterpretation of key concepts of participation, gender and empowerment in ways that diluted their original purpose and changed their impact. Influential development agencies and donors also appropriated these concepts. Participatory research and gender mainstreaming were rationalized in international agricultural research bureaucracies by their potential for improving research efficiency, for reaching more farmers and increasing technology adoption. For the most part, from 1980 to 2020, gender-intentional participatory research fulfilled this promise. However, this was achieved at a price: a divorce from action research and a failure to build an organized client base that could both communicate demand for new technology and hold research providers accountable to farmers. Weak linkages to demand from an organized client-base of farmers enabled science bureaucracies to defuse the power reversals that could occur through transformative change in gender relations, and if men and women farmers had gained agency through participation in research.

Transformative approaches often require intensive, local support. When supported by effective second and third tier collective organization, such as achieved by women's sorghum groups in Mali (Rattunde et al. 2021) rice-wheat groups in India (Isaacs et al. 2024), bean farmers in Honduras (Humphries et al. 2012) and farmer field schools (Van den Berg et al. 2020), these have scaled out to reach thousands of farmers. The trajectory of empowerment was different in NGOs committed to action research and collective action, from solid organizations where farmers were already able to exert effective demand on research. Perhaps science-led, gender-intentional participatory research faces a trade-off between transformative impacts and scale, but difficulty in scaling up is less an inherent constraint than the result of poor linkages to second and third tier collective organization of their client base.

This reflection on past experience suggests that transformative gender-intentional participatory research in agriculture requires a recommitment to the original action research concept of participatory research. This will require an inclusive approach that goes beyond simple sex dichotomies, to renew principles of respecting farmers' knowledge, strengthening farmer-led experimentation and reversing the power balance in scientist-farmer relationships. Action research is an extensive field, well-established as a policy instrument in the health and education sectors, with a wealth of experience to draw on (Bradbury 2015; Somekh and Noffke 2009). But action-research alone will not be enough.

Science bureaucracies must also be proactive partners with farmer organizations, women's collectives and supporting NGOs in transformative efforts to overcome structural causes of inequality and poverty. This means investment in the co-production of technology involving stakeholders of many types, finding ways to blend mainstream science with local knowledge and to mobilize large-scale citizen science. This will require a more sophisticated understanding of client demand for innovation than is presently used. Outcomes of co-produced research in agriculture should include enhancing rural men's and women's own capacity for innovation. Empowerment should take on a broader meaning, beyond individual farmers' social relations in their households, communities, and value chains to include having voice and agency in technical innovation. To achieve this, gender-intentional participatory research aiming for large-scale transformative impact on agriculture requires a fundamental change of approach to engagement with its client base among the rural poor.

References

Abrol D, Gupta A (2014) Understanding the diffusion modes of grassroots innovations in India: a study of Honey Bee Network supported innovators. Afr J Sci Technol Innov Dev 6:541–552. https://doi.org/10.1080/20421338.2014.976974

Acosta M, van Bommel S, van Wessel M, Ampaire EL, Jassogne L, Feindt PH (2019) Discursive translations of gender mainstreaming norms: the case of agricultural and climate change policies in Uganda. Women's Stud Int Forum 74:9–19

Almekinders CJM, Elings A (2001) Collaboration of farmers and breeders: participatory crop improvement in perspective. Euphytica 122:425–438

Almekinders CJM, Hardon J, Christinck A, Humphries S, Pelegrina D, Sthapit B, Vernooy R, Visser L, Weltzien E (2006) Bringing farmers back into breeding. Experiences with participatory plant breeding and challenges for Institutionalisation. Agromisa Special 5, Agromisa, Wageningen. pp 135

Altieri MA (1989) Agroecology: a new research and development paradigm for world agriculture. Agric Ecosyst Environ 27(1–4):37–46

Altieri MA, Nicholls CI (2017) Agroecology: a brief account of its origins and currents of thought in Latin America. Agroecol Sustain Food Syst 41(3–4):231–237

Altieri MA, Toledo VM (2011) The agroecological revolution in Latin America—rescuing nature, ensuring food sovereignty and empowering peasants. Peasant Stud 38:587–612

Alvarez S, Staiger-Rivas S, Tehelen K, García CX, Manners G, Biermayr-Jenzano P (2010) Demand analysis report: gender-responsive participatory research. Centro Internacional de Agricultura Tropical (CIAT), Cali, Working document no. 215

Ashby J (2009) Fostering farmer first methodological innovation: organizational learning and change in international agricultural research. In: Scoones I, Thompson J (eds) Farmer first revisited: innovation for agricultural research and development. Practical Action Publication, London

Ashby JA, Braun AR, Gracia T, Guerrero MDP, Hernández Romero LA, Quirós Torres CA, Roa Velasco JI (2000) Investing in farmers as researchers: experience with local agricultural research committees in Latin America. Cali. Colombia: CIAT

Batliwala S (2002) Taking the power out of empowerment–an experiential account. Dev Pract 17(4–5):557–565

Bebbington A, Farrington J, Lewis DJ, Wellard K (2005) Reluctant partners? Non-governmental organizations, the state and sustainable agricultural development. Routledge, New York

Benítez Fernández B, Nelson E, Crespo Morales A, Ortiz Pérez R, Acosta Roca R, Cárdenas Travieso RM (2023) Transforming food systems: a gendered perspective on local agricultural innovation in Cuba. Front Sociol 8:1256379

Bidwell D (2009) Is community-based participatory research postnormal science? Sci Technol Hum Values 34(6):741–761

Biggs SD (1989) Resource-poor farmer participation in research: a synthesis of experiences from nine national agricultural research systems. OFCOR Series (Comparative Study)

Blau PM, Meyer MW (1971) Bureaucracy in modern society, 2nd edn. Random House, New York

Bradbury H (2015) The Sage handbook of action research. Sage, Los Angeles

Brummett RE, Jamu DM (2011) From researcher to farmer: partnerships in integrated aquaculture–agriculture systems in Malawi, Ghana and Cameroon. Int J Agric Sustain 9(1):282–289

Burns D, Howard J, Ospina S (2021) Challenges in the practices of participatory research and inquiry. In: Burns D, Howard J, Ospina S (eds) The SAGE handbook of participatory research and inquiry, 17. SAGE, London

Ceccarelli S, Grando S (2019) From participatory to evolutionary plant breeding. In: Farmers and plant breeding: current approaches and perspectives. Routledge, London, p 231

Ceccarelli S, Grando S (2020) Participatory plant breeding: who did it, who does it and where? Exp Agric 56:1–11

CGIAR (1995) Report on the CGIAR gender program. CGIAR Secretariat Document No ICW/95/07, Washington, DC

CGIAR Initiative on Market Intelligence (2024) CGIAR research initiative on market intelligence: annual technical report 2023. CGIAR System Organization, Montpellier. https://hdl.handle.net/10568/141612

Chambers R (1997) Whose reality counts? Putting the first last. IT Publications, London

Chambers R, Thrupp LA (1994) Farmer first: farmer innovation and agricultural research. Karthala Editions, Paris

Christinck A, Diarra M, Hornber G (2014) Innovations in seed systems: lessons from the CCRP funded project "sustaining farmer-managed seed initiatives in Mali, Niger, and Burkina Faso". The McKnight Foundation, Minneapolis. https://www.ccrp.org

Clisby S, Enderstein AM (2017) Caught between the orientalist–occidentalist polemic: gender mainstreaming as feminist transformation or neocolonial subversion? Int Fem J Polit 19(2):231–246

Cooke B, Kothari U (eds) (2001) Participation: the new tyranny? Zed Books, New York

Cornwall A (2003) Whose voices? Whose choices? Reflections on gender and participatory development. World Dev 31(8):1325–1342

Cornwall A, Brock K (2005) What do buzzwords do for development policy? A critical look at "participation", "empowerment" and "poverty reduction". Third World Q 26(7):1043–1060

Cornwall A, Rivas AM (2015) From "gender equality" and "women's empowerment" to global justice: reclaiming a transformative agenda for gender and development. Third World Q 36(2):396–415

Cleaver F (2017) Development through bricolage: rethinking institutions for natural resource management. Routledge, London

de Boef WS, Singh S, Trivedi P, Yadav KS, Mohanan PS, Kumar S, Yadavendra JP, Isaacs K (2021) Unleashing the social capital of self-help groups for strengthening seed systems in Uttar Pradesh, India. Glob Food Secur 29:100522. https://doi.org/10.1016/j.gfs.2021.100522

De Jong S, Kimm S (2017) The co-optation of feminisms: a research agenda. Int Fem J Polit 19(2):185–200

Dewulf A, Craps M, Bouwen R, Abril F, Zhingri M (2005) How indigenous farmers and university engineers create actionable knowledge for sustainable irrigation. Action Res 3(2):175–192

Eade D, Friis-Hansen E (2008) Impact assessment of farmer institutional development and agricultural change: Soroti district, Uganda. https://cgspace.cgiar.org/bitstream/handle/10568/76152/Friis-Hansen.pdf?sequence=1

Fals-Borda O (1987) The application of participatory action research in Latin America. Int Sociol 2(4):329–347

Fals Borda O (1998) People's participation challenges ahead. Intermediate Technology Publications, London

FAO (2018) The future of food and agriculture – alternative pathways to 2050. Rome. 224 pp. Licence: CC BY-NC-SA 3.0 IGO

FAO (2011) The state of food and agriculture, 2010–2011. Food and Agriculture Organization, Rome

FAO (2023) The status of women in agrifood systems. Food and Agriculture Organization, Rome. https://doi.org/10.4060/cc5343en

FAO (Food and Agriculture Organization of the United Nations), IFAD (International Fund for Agricultural Development), UNICEF, WFP (World Food Programme), WHO (World Health Organization) (2022) The state of food security and nutrition in the world 2022. Repurposing food and agricultural policies to make healthy diets more affordable. FAO, Rome

Feldstein HS (2000) Gender analysis: making women visible and improving social analysis. In: Collinson M (ed) A history of farming systems research. FAO and CABI, Rome, p 67

Freire P (1970) Pedagogy of the oppressed. Ramos MB (Trans). Continuum, New York

Gaventa J, Cornwall A (2008) Power and knowledge. In: The Sage handbook of action research. Participative inquiry and practice, 2nd edn. SAGE Publications, London, pp 172–189

Gramsci A (1971) In: Hoare Q, Smith GN (eds) Selections from the prison notebooks. Lawrence and Wishart, London

Gliessman SR, Rosemeyer M (2009) The conversion to sustainable agriculture: principles, processes, and practices. CRC Press, Boca Raton

Harris-Fry H, Nur H, Shankar B, Zanello G, Srinivasan C, Kadiyala S (2020) The impact of gender equity in agriculture on nutritional status, diets, and household food security: a mixed-methods systematic review. BMJ Glob Health 5(3):e002173

Herdt RW, Anderson JR (2019) The contribution of the CGIAR centers to world agricultural research. In: Policy for agricultural research. CRC Press, New York, pp 39–64

Humphries S, Classen L, Jiménez J, Sierra F, Gallardo O, Gómez M (2012) Opening cracks for the transgression of social boundaries: an evaluation of the gender impacts of farmer research teams in Honduras. World Dev 40(10):2078–2095

Isaacs K, Weltzien E, Some H, Diallo A, Diallo B, Sidibé M, Vom Brocke K, Samake B, Nebié B, Rattunde FW (2024) Increasing sorghum yields for smallholder farmers in Mali: the evolution towards a context-driven, on-farm, gender-responsive sorghum breeding program. Front Sustain Food Syst 8:1334385

Jackson C, Pearson R (eds) (2005) Feminist visions of development: gender analysis and policy. Routledge, New York

Jasanoff S (2002) New modernities: reimagining science, technology and development. Environ Values 11(3):253–276

Jiggins J (1986) Gender-related impacts and the work of the international agricultural research centers. CGIAR Secretariat, Washington, DC

Johnson NL, Lilja N, Ashby JA (2003) Measuring the impact of user participation in agricultural and natural resource management research. Agric Syst 78(2):287–306

Kimura AH, Kinchy A (2016) Citizen science: probing the virtues and contexts of participatory research. Engag Sci Technol Soc 2:331–361

Kothari U (2001) Power, knowledge and social control in participatory development. In: Cooke B, Kothari U (eds) Participation: the new tyranny? Zed Books, New York

Laajaj R, Macours K, Masso C, Thuita M, Vanlauwe B (2020) Reconciling yield gains in agronomic trials with returns under African smallholder conditions. Sci Rep 10:14286. https://doi.org/10.1038/s41598-020-71155-y

Louis M, Maertens L (2021) Why international organizations hate politics: depoliticizing the world. Taylor & Francis, Milton

Leal PA (2010) Participation: the ascendancy of a buzzword in the neo-liberal era. In: Cornwall A, Eade D (eds) Deconstructing development discourse, buzz words and fuzz words. Oxfam, Warkwickshire, pp 89–101

Lengwiler M (2008) Participatory approaches in science and technology: historical origins and current practices in critical perspective. Sci Technol Hum Values 33(2):186–200

Lilja N, Ashby JA (1999) Types of participatory research based on locus of decision making. CGIAR Systemwide Program on Participatory Research and Gender, Working Document no. 6, Cali

Lilja N, Bellon M (2008) Some common questions about participatory research: a review of the literature. Dev Pract 18(4–5):479–488

Lykes MB, Coquillon E (2007) Participatory action research and feminisms. In: Handbook of feminist research: theory and praxis. Sage, Thousand Oaks, pp 297–326

Miller C, Razavi S (2011) Missionaries and mandarins: feminist engagement with development institutions. In: Staab S, Razavi S (eds) Gendered dimensions of development. United Nations Research Institute for Social Development (UNRISD), Geneva, pp 311–329

Moser C (2005) Has gender mainstreaming failed? Int Fem J Polit 7(4):576590. https://doi.org/10.1080/14616740500284573. https://doi.org/10.1080/13552070512331332283

Moser C, Moser A (2005) Gender mainstreaming since Beijing: a review of success and limitations in international institutions. Gend Dev 13(2):11–22

Mosse D (1994) Authority, gender and knowledge: theoretical reflections on the practice of participatory rural appraisal. Dev Chang 25(3):497–526

Mosse D (2003) The making and marketing of participatory development. In: A moral critique of development. Routledge, Mirano, pp 57–89

Moreira M, Mulvany P (2009) Farmers' movements and the struggle for food sovereignty in Latin America. In: Scoones I, Thompson J (eds) Farmer first revisited: innovation for agricultural research and development. Practical Action Publication, London

Narayan D, Petesch P (eds) (1997) Voices of the poor from many lands. World Bank, Washington, DC

Njuki J, Kaler A, Parkins JR (2016) Conclusion: towards gender transformative agriculture and food systems: where next? In: Njuki J, Parkins JR, Kaler A (eds) Transforming gender and food security in the global south. Routledge, New York, pp 283–291

Paris TR, Singh A, Cueno AD, Singh VN (2008) Assessing the impact of participatory research in rice breeding on women farmers: a case study in Eastern Uttar Pradesh. India. Exp Agric 44. https://doi.org/10.1017/S0014479707005923

Poats S (1990) Gender issues in the CGIAR system: lessons and strategies from within. CGIAR, Rome

Pohl C, Rist S, Zimmermann A, Fry P, Gurung GS, Schneider F, Speranza CI, Kiteme B, Boillat S, Serrano E, Hadorn GH (2010) Researchers' roles in knowledge co-production: experience from sustainability research in Kenya, Switzerland, Bolivia and Nepal. Sci Public Policy 37(4):267–281

Pretty JN (1995) Participatory learning for sustainable agriculture. World Dev 23(8):1247–1263

Pretty J, Attwood S, Bawden R, Van Den Berg H, Bharucha ZP, Dixon J, Flora CB, Gallagher K, Genskow K, Hartley SE (2020) Assessment of the growth in social groups for sustainable agriculture and land management. Glob Sustain 3:e23

Ravetz JR, Funtowicz S (2001) New forms of science. In: Smelser N, Baltes P (eds) International encyclopedia of the social and behavioral sciences. Elsevier, Amsterdam, pp 13683–13687

Rattunde F, Weltzien E, Sidibé M, Diallo A, Diallo B, Vom Brocke K, Nebié B, Touré A, Traoré Y, Sidibé A, Diallo C (2021) Transforming a traditional commons-based seed system through collaborative networks of farmer seed-cooperatives and public breeding programs: the case of sorghum in Mali. Agric Hum Values 38(2):561–578

Resurrección BP, Elmhirst R (2020) Lifting the barriers of gender integration in livestock production. In: Resurrección BP, Elmhirst R (eds) Negotiating gender expertise in environment and development. Routledge, New York, pp 171–183

Rhoades RE, Booth RH (1982) Farmer-back-to-farmer: a model for generating acceptable agricultural technology. Agric Adm 11:127–137

Richards P (1985) Indigenous agricultural revolution: ecology and food production in West Africa. Hutchinson Education, London

Rist S, Chiddambaranathan M, Escobar C, Wiesmann U (2006) "It was hard to come to mutual understanding…"—the multidimensionality of social learning processes concerned with sustainable natural resource use in India, Africa and Latin America. Syst Pract Action Res 19:219–237

Rocheleau D, Thomas-Slayter B, Wangari E (1996) Feminist political ecology: global issues and local experiences. Zed Books, New York

Roth WD, Sonnert G (2011) The costs and benefits of "red tape": anti-bureaucratic structure and gender inequity in a science research organization. Soc Stud Sci 41(3):385–409

Rosset PM, Martínez-Torres ME (2014) Food sovereignty and agroecology in the convergence of rural social movements. In: Alternative agrifood movements: patterns of convergence and divergence. Emerald Group Publishing Limited, Bradford, pp 137–157

Rosario BP (2009) Fostering farmer–scientist research collaboration: the role of the International Federation of Agricultural Producers. In: Scoones I, Thompson J (eds) Farmer first revisited: innovation for agricultural research and development. Practical Action Publication, London

Scoones I, Thompson J (eds) (2009) Farmer first revisited: innovation for agricultural research and development. Practical Action Publication, London

Sen A (1988) The concept of development. In: Handbook of development economics, Volume 1. Elsevier, pp 9–26

Somekh B, Noffke SE (2009) The SAGE handbook of educational action research. Sage, London

Sumberg J, Okali C (1997) Farmers' experiments: creating local knowledge. Lynne Rienner Publishers, Boulder

Sumberg J, Thompson J, Woodhouse P (2013) Why agronomy in the developing world has become contentious. Agric Hum Values 30:71–83. https://doi.org/10.1007/s10460-012-9376-8

Tarjem IA, Westengen OT, Wisborg P, Glaab K (2022) "Whose demand?" The co-construction of markets, demand and gender in development-oriented crop breeding. Agric Hum Values 40:83–100. https://doi.org/10.1007/s10460-022-10337-y

Tavenner K, Crane TA (2022) Hitting the target and missing the point? On the risks of measuring women's empowerment in agricultural development. Agric Hum Values 39:849–857

Tribaldos T, Oberlack C, Schneider F (2020) Impact through participatory research approaches: an archetype analysis. Ecol Soc 25. https://doi.org/10.5751/ES-11517-250315

Van den Berg H, Phillips S, Dicke M, Fredrix M (2020) Impacts of farmer field schools in the human, social ,natural and financial domain :a qualitative review. Food Secur 12(6):1443–1459. https://doi.org/10.1007/s12571-020-01046-7

van der Burg M (2019) "Change in the making": 1970s and 1980s building stones to gender integration in CGIAR agricultural research. In Sachs, C.E., (ed). 2019. Gender, Agriculture and Agrarian Transformations (pp. 3-7). Routledge. https://doi.org/10.4324/9780429427381

Vom Brocke K, Kondombo CP, Guillet M, Kaboré R, Sidibé A, Temple L, Trouche G (2020) Impact of participatory sorghum breeding in Burkina Faso. Agric Syst 180:102775

Walker TS (2006) Participatory varietal selection, participatory plant breeding, and varietal change. World Bank, Washington, DC

Waters-Bayer A, Wettasinha C, van Veldhuizen L (2009) Building partnerships to promote local innovation processes. In: Scoones I, Thompson J (eds) Farmer first revisited: innovation for agricultural research and development. Practical Action Publication, London

Waters-Bayer A, Kristjanson P, Wettasinha C, van Veldhuizen L, Quiroga G, Swaans K, Douthwaite B (2015) Exploring the impact of farmer-led research supported by civil society organisations. Agric Food Secur 4(1):1–7

WIRF (1991) From field to lab and back. Women and rice farming systems. Women and Rice Farming Systems program. https://cgspace.cgiar.org/handle/10947/5702

World Bank (2007) World development report 2008: agriculture for development. The World Bank, Washington, DC. http://hdl.handle.net/10986/5990

World Bank (2012) World development report, 2012: gender equality. The World Bank, Washington DC

Chapter 5
The Early Embedding of Hegemonic Masculinity in International AR4D: Norman Borlaug, His Ways of Working and Their Gendered Implications

Franz F. Wong ⓘ

Dedicated to the memory of Margaret Borlaug

Abstract This chapter aims to better understand how agricultural research for development (AR4D) has been historically constituted as gendered practice by tracing early influences on wheat agronomy research. I develop a historical account of Norman Borlaug and his agronomic innovations to make visible early gendered research practices. While these were common during the time of Borlaug's seminal

---

In CIMMYT, the ideal worker is one who will travel great distances and work long hours, even at the expense of family life.

(Merrill-Sands et al. 1999)

I ended up leaving CIMMYT because I got pregnant … just the pressures of the fieldwork and everything else. I decided that it was probably safer for me to quit.
(former CIMMYT researcher 2015; CIMMYT is the Spanish acronym for the "The International Maize and Wheat Improvement Center" based in Mexico. While the focus of this chapter is not specifically on CIMMYT, many of the examples are drawn from this research organization, founded in 1966, given its long association with Norman Borlaug. As the main international center for wheat research, it remains strongly associated with his legacy.)

This chapter was made possible with the financial support from the Global Center for Gender Equality. Thanks to Sarah Henry for her belief in the adventure. I have tremendously benefited from discussions with and inputs from Lone Badstue as well as the editors and anonymous reviewers of this book. Also thanks to Gary Darmstadt, Marci Baranski, Chris Hunter, Margreet van der Burg, and Diana Lopez for their insights as well as CIMMYT staff for providing access to its training database. I would like to also acknowledge David Ludwig for his observation of Norman Borlaug hagiography.

F. F. Wong (✉)
Independent Consultant, Amsterdam, Netherlands
e-mail: fwong@aspireinternational.ca

© The Author(s) 2025   71
J. Njuki et al. (eds.), *Gender, Power and Politics in Agriculture*,
https://doi.org/10.1007/978-3-031-60986-2_5

work in the 1950s, I explore how they have, in their essence, been reproduced as hegemonic masculine practice. In particular, the mobilization of the "ideal" wheat researcher, through repetitive, stylized acts such as fieldwork, are concerned with gender performativity. Specific ways of working then get reproduced through professional development and career advancement opportunities and served as a basis for inclusion and exclusion of what it is to be a wheat researcher. They are gendered in that it is generally men who could avail themselves of such opportunities as they were generally free of social reproduction activities, yet benefited from them. Conversely, those who undertook such care work, mainly women, have been more challenged to succeed as agronomists within the prevailing work paradigm. I also suggest that upholding such masculinized norms of work can have adverse health and personal well-being implications for men and others who pursue such practices.

5.1   Introduction

Gender mainstreaming efforts in agricultural research for development (AR4D) have mainly focused on making the case for the integration of gender issues. However, few researchers (except for Mukhopadhyay and Prügl 2019) have explored: what is the mainstream that gender is being integrated into? Given the long history of addressing gender issues in agricultural development more generally, a more appropriate question is: *How has AR4D historically been constituted that has made change and gender integration particularly difficult?* By understanding how the structure of knowledge-making in agricultural research has historically been gendered, we can start to understand how to address institutional change, politically, normatively, and in practice (Pyburn and van Eerdewijk 2021).

This chapter is a gender analysis, drawing on feminist technoscience studies and hegemonic masculinities, to explore how masculinity became constitutive of AR4D and how AR4D practice further reproduces its own gendered-ness. The male domination of agricultural development more generally, as well as the male bias of AR4D in particular, is well documented (Crewe and Harrison 1998; Farhall and Rickards 2020; Pyburn and van Eerdewijk 2021; Sachs et al. 2021). Using the example of wheat agronomy in the early days of international AR4D, this chapter explores how knowledge generation practices are historically gendered and foundational to the maintenance of male-dominated agricultural research. In other words, AR4D not only reproduces gender bias but performs gender as a knowledge enterprise (Mukhopadhyay and Prügl 2019).

The focus of this chapter is on Norman Borlaug and his influence on wheat breeding. The 1970 Nobel Peace Prize recipient and other scientists developed agricultural innovations for what has become known as the "Green Revolution", in response to the devastating famines in South Asia when millions died in the 1960s. There is a plethora of research on what was and was not achieved during the Green Revolution. Its social, economic and political trans-global impacts on agriculture as well as on participating communities, farming households and their members remain controversial. There are also numerous accounts of Borlaug's professional and, to a lesser degree, personal life and his contributions to addressing global hunger as well as agricultural research, particularly agronomy and plant breeding. A glance at the monikers used in articles, dedications and biographies reveals how the researcher is venerated. These include "Dedication: Norman E. Borlaug The Humanitarian Plant Scientist Who Changed the World" (Ortiz and Mowbray 2007), "The Man Who Fed the World" (Hesser 2009), "Father of the Green Revolution" (Borlaug 2008 and Tomar 2009 cited by Sumberg et al. 2012), "The Man Who Saved a Billion Lives", and the one "who proved Malthus wrong" (Tuns 2009).

Rather than enter the dragon's den of the claims and critiques of the Green Revolution, I focus on Borlaug's ethos to his work and related gendered patterns of his work practices as both a researcher and as a teacher. I argue that what his privileged in knowledge production informed what it is to be an agricultural researcher and who is able succeed as such. Through his own work, particularly by way of demonstration and fellowship, as well as through his ubiquitous teaching and training, such practices and ways of working proliferated and became institutionalized through professional development and practice in wheat agronomy.

As Goetz (1997, p. 17) writes, male dominance is due to "the historical embedding of (men's) needs and interests in the structures and practices of public institutions". My aim is to develop a historical, gendered account about Borlaug to make visible dimensions of his work previously obscured. To paraphrase the editors of the *Gender and History* journal, in this chapter I center gender relations in the study of AR4D and bring a historical perspective to the study of gender in AR4D (The Editorial Collective 1989).

5.2 Methodology and Background

As I set out to learn and write about such a male icon as Norman Borlaug, I soon realized two methodological complications. The first arises from working with historical data; how do I avoid imposing a normative lens on 80-year old events by writing with a contemporary critical perspective? Understanding context seems key: gender relations in the U.S. during the 1940s and 1950s, the focus of much of this chapter, can be broadly characterized as "traditional" with married women generally assuming household roles and men as 'breadwinners". This was also a time of change during and after World War II when gender roles were disrupted while, at the same time, heroic and hyper-masculine images of men dominated popular culture in the U.S., e.g., the Marlboro Man, and John Wayne. Accordingly, Borlaug's gendered values and male predominance, numerically or otherwise, described in this chapter are not particularly unique to him and his situation. My purpose of highlighting these is not to pass judgement on a time gone by, but to acknowledge historic gender practice as foundational. And while I focus on one man, my aim is not to attribute all AR4D practice to him, but to highlight a major influence.

The second challenge concerns focusing on a venerated man and his accomplishments, in the name of gender equality, as a man myself. This is complicated by the nature of the sources I have used that exclusively, with a few exceptions, come from men, many of whom were Borlaug's peers. How do I write critically without reinscribing the very male-biased accounts that have preceded me (Ditz 2004)?

Connell (1993, p. 284) cautions about writing history and masculinity without any political positioning. In arguing for conceptual framing to address "an increasingly reactionary celebration of masculinity", she suggests a focus on men where masculinity is not the main frame. Accordingly, my aim is to situate the analysis within a feminist critique of science and an understanding of gender as performance of hegemonic masculinity in order to generate additional insights to what otherwise has mostly been a hagiography of a significant historical figure.

5.2.1 Feminist Critique of Science and Hegemonic Masculinities

Feminist scholars have long challenged what Haraway (1988) has referred to as the "god trick": the shroud that provides long-held claims of "objective", "neutral" and "rational" science (Sachs et al. 2021). This ultimately obscures what lies beneath: science, in all its dimensions, reflecting particular knowledges derived from particular ways of knowing. These specificities not only mirror, albeit with distortions, the contexts in which knowledge is generated but also that of the knowledge makers: the scientists and their paradigms (Åsberg and Lum 2010). In other words, science is in reality value-laden.

To be sure, the case is not that scientists necessarily claim a position of complete neutrality. Norman Borlaug was unabashedly upfront of his agendas: to save the world from global hunger, to influence national food and agricultural policies and to affect how agricultural research was conducted. It is the latter that is the focus of this chapter: I am not challenging Borlaug's agronomic findings. Rather, I am interested in understanding how gender is structured in agricultural research and how the gender order is maintained under the guise of assumed objective science due to its "rational" basis.

Feminist technoscience studies provides a basis for a critical questioning of "scientific practices, about researchers' positioning, about consequences of these practices for different actors, and about power issues related to knowledge-making and scientific practices" (Sefyrin et al. 2018, p. 2). In the context of transglobal AR4D, feminist technoscience studies are explored for their potential to illuminate the reproduction of male bias. In AR4D, this not only refers to the continued numerical domination of men scientists, bias also concerns a certain way of thinking about and undertaking research that maintains authority and privilege of generally white, cisgender men scientists and the subordinate positions of scientists not associated with dominant intersectional identities of gender, sexual orientation and race. I am not just referring, for example, to active discrimination of women researchers, but also to how gender is performed in ways that create exclusive ways of knowledge-generation and formal and informal practices of exclusion and domination, many of which are subtle (Page et al. 2009). Epistemically, this concerns what are considered valid research topics, evidence and ways of generating such evidence.[1]

Moreover, this chapter draws on notions of hegemonic masculinity (Connell and Messerschmidt 2005) as a way to understand the predominance of men in mainstream AR4D and related male-centric methodologies. I mainly draw on engineering and masculinity literature, given the relative dearth of research on masculinities in AR4D, particularly in the Global South (Tickamyer and Sexsmith 2019). I also consider the analysis of the gendered category of men (Farhall and Rickards 2020) to understand how gender is reproduced in and through organizational processes and social relations (Rao et al. 1999).

We can understand masculinity as performance: "practices and ways of being that serve to validate the masculine subject's sense of himself as male/boy/man" (Whitehead 2002, p. 5 cited by Liebrand 2014). Gender is performative as stylized acts: repeated performance that is "at once a reenactment and reexperiencing of a set of meanings already socially established … (constituting the) mundane and ritualized form of their legitimation" (Butler 1988, p. 526).

In engineering, the repetition of ritualized, albeit diverse, performances of masculinity become unquestioned, particular ways-of-being and doing and "increasingly associated with 'men' and building structures" (Liebrand and Udas 2017, pp. 5–6). For example, within the engineering university setting, "engineering students are trained to do science and technology in specific ways that embody particularly masculine symbolic repertoires" (Rap and Oré 2017, p. 95). The repeated acts of performance, through "tacit collective agreement", mean the performers become subsumed by their own makings of reality, making its reproduction critical to the sustaining of a sense of natural-ness (Vera-Gajardo 2021, p. 7).

Part of this tacit agreement concerns particular ways of being a man that subsume others, resulting in a hierarchy among men as well as with those of other genders where race and class are also in play. Connell and Messerschmidt (2005) describe such gender relations as hegemonic masculinity in recognition of multiple ways of being masculine, some of which are the standards against which others are deemed subordinate. In using the term "hegemonic", Connell and Messerschmidt draw in Gramsci's notion of domination through both consent and coercion.

Within a particular sector, how does a particular expression of masculinity become hegemonic? Liebrand (2014 citing Faulkner 2009, p. 15) suggests that "communities of practice" are both stage and audience where "actors" learn what is considered "normal" by their participation, thereby learning "gender-authentic" performance. A community of practice is a "group of people gather(ing) around a common commitment or purpose" where gender-performative practices "appear in

---

[1] "Male bias" also concerns bias in the benefits derived from biased AR4D. With specific research topics privileged for specific end-users, the notion that men are farmers is reproduced and other farmers, particularly women and women of particular classes, ethnicities and civil status, are ill-considered if not left out of the mainstream AR4D altogether (Barbarcheck 2021).

the evolution of this shared purpose" (Faulkner 2009, p. 3). Integral is a sense of belonging that "originates from the stories people tell one another and themselves about who they are (and who they are not)" (Faulkner 2009, p. 8). These become the basis of identification and emotional attachments while also having a delineation effect: they contribute to a yearning to belong as well as the power to determine who belongs.

Still, the frame of masculinities can produce particular ambiguities, particularly when not understood as a social relation and understood contextually. Liebrand (2014, p. 17) suggests that the study of masculinities "should be as specific as possible about performances of masculinity … and the perspective from which it is done" in acknowledgement that performing masculinity likely says more about the "do-er" than the context per se. Hence my focus on an individual in this chapter within the wider context of AR4D.

5.2.2   Working with Historic Biographical Data

Given that primary data collection from Borlaug is not possible, secondary sources are the only ones available to understand the past. While Borlaug was a prolific writer, his catalogue includes little biographical data except for interviews (for example, Borlaug 2008). While the hagiographic literature about Borlaug was written from particular perspectives, their biases can be uncovered by a gendered analysis. For it is the "collection of values, history, culture and practices that form the unquestioned, 'normal' way of working in organizations" (Rao et al. 1999, p. 2). In particular, understanding the history of AR4D, or at least a seminal part of it, can "illuminate how patterns of exclusion became institutionalized" (Goetz 1997, p. 16) and requires new interpretations of history to better understand taken for granted facts and their construction.

Much of the biographical data used in this chapter are from two main authors, namely Noel Vietmeyer (2011a, b) and Leon Hesser (2009). Their work also served as main sources for other writings about Borlaug (Mann 2018).[2] These biographers have had a long history with him. Having worked and travelled with Borlaug as chair of several review panels of the US National Academy of Science, Vietmeyer collected some "300 personal experiences" from Borlaug. These "make up the heart and soul" (Vietmeyer 2011a, p. 302) of Vietmeyer's several biographies on Borlaug, whom Vietmeyer refers to as the "father of the food supply" (Vietmeyer 2011a, p. 302). As member of the US Foreign Service in Pakistan, Hesser worked with Borlaug from 1966 when he headed the team that introduced Borlaug's technologies. Later in 1973, he oversaw the US support for the establishment of AR4D

---

[2] For example, Mann (2018) draws heavily from Vietmeyer (2011a) and considers his work the closest to a Borlaug auto-biography given Borlaug's extensive editorial control.

centers that would later comprise the Consultative Group on International Agricultural Research (CGIAR). In addition, I have drawn from other accounts of Borlaug's work by his peers, particularly eulogies.

I privilege these sources of biographical information for two reasons. First, they have been endorsed by Borlaug, as in the case of Vietmeyer (2011a, p. 269), where Borlaug embraces their factual basis and authenticity of his account. Second, their own essence and what they emphasize is treated as data for the purpose of this chapter and its analysis. Accordingly, these sources are quoted extensively, drawing on discursive strategies more aligned with ethnographic approaches to writing than, for example, writing for bio-physical publications. In addition, I have also drawn from biographical information and critique from the few sources that have steered away from hagiography, namely Mann (2018), Sumberg et al. (2012), Baranski (2022) and Laveaga (2021).

5.2.3   Norman Borlaug—Biographical Sketch

Borlaug was born in 1914 on his grandparent's farm in rural Iowa, USA. He grew up with his parents and two sisters in the nearby Norwegian community of Saude. He attended a one-room primary school and later graduated in forestry from the University of Minnesota in 1937, the first of his family to earn a university degree. His family lived a hard life as farmers in rural US, particularly during the Great Depression. Despite this, Borlaug was deeply affected by the poverty he observed when he first went to Minnesota for university, where he witnessed food riots. After a stint as an assistant forest ranger in 1938, he returned to the University of Minnesota where he earned a PhD in plant pathology in 1941.

After graduating he worked for a large chemical company, DuPont, as a plant pathologist from 1941 to 1944, during World War II. In October 1944, Borlaug joined the Cooperative Wheat Research and Production Program of the Office of Special Studies, a joint program of the Mexican Government and the Rockefeller Foundation. During this time in Mexico Borlaug developed seminal breeding techniques, such as "shuttle breeding" and high-volume cross breeding,[3] and semi-dwarf wheat varieties. These innovations contributed directly to several countries achieving self-sufficiency in food production such as Pakistan (1968) and India (1974).

---

[3] Shuttle breeding entailed growing lines of wheat in two distinct parts of Mexico, 3500 kilometers apart, allowing two crops a year, thereby halving the time required to breed new wheat varieties. It was called "shuttle breeding" because Borlaug and his fellow breeders shuttled every year between research stations. High-volume crossbreeding accelerated plant breeding by working with thousands of varieties of wheat. This otherwise "hit-and-miss" (Hesser 2009, p. 44) process was made more efficient by Borlaug, including sheer numbers of plants that needed to be crossed during pollination and in the field. For more detailed descriptions of Borlaug's innovations, see Hesser (2009), van Ginkel et al. (2002), and Mann (2018).

Borlaug later went on to become of the first Director of the International Wheat Improvement Program in 1966 at CIMMYT until 1979, after which he served as a consultant for many years (Sumberg et al. 2012). CIMMYT and its present flagship Wheat Program descend from the Cooperative Wheat Research and Production Program of the Office of Special Studies (van Ginkel et al. 2002).

Of the numerous awards Borlaug received, the most notable are the First International Services Award in Agronomy (1968), the Nobel Peace Prize (1970), the President's Medal of Freedom (1977 in the US), and the World Food Prize (1996). Borlaug passed away in 2009 at 95-years-old.

5.3   Embedding of Hegemonic Masculinities in Wheat Agronomy

This section uses a feminist technoscience lens to analyze Borlaug's ways of working, where I foreground what he believed to be important in wheat breeding and what his peers found as critical to the work of agronomy. I then explore their gendered implications.

5.3.1   Heroic Individualism

Rao et al. (1999) identify "heroic individualism" as one of four dimensions of the deep structures of organizations that perpetuate gender inequality. It is the myth of a crusader that not only encourages a "culture of winning" but also gendered organizations where men are privileged and women are disadvantaged. A reading of descriptions of Borlaug's work point to a "heroic individualism" theme in his work ethos and ways of working: a singularly focused and seemingly infinite determination to reduce world hunger by increasing yields through technological innovations. Sumberg et al. (2012, p. 1591) refer to his "single-mindedness and a strong work ethic" as his defining trait portrayed by biographers. In the following I explore the basis of the portrayal of Borlaug as a crusader.

Borlaug, whom Vietmeyer (2011b, p. 97) refers to as the "Lone Ranger of wheat research", often experienced adverse working conditions, particularly during his early research days in Mexico. The work entailed long hours, working from or before sunrise to past sunset, often seven days a week, motivated, as he recalls, by the hope of creating "better wheats for farmers" (Vietmeyer (2011b, p. 98).[4] The physical conditions in Chapingo, in central Mexico, were harsh with relentless heat

---

[4] Later, in his dedication to Borlaug, Rajaram (2011) writes that "Unlike most other mortals … the notion of weekend was an alien concept to him". Hired by Borlaug, Rajaram worked at CIMMYT from 1969 to 2003, and was its Director of the Global Wheat Program from 1996 to 2003. He won the World Food Prize in 2014.

and cold evenings while the living conditions were basic. When Borlaug shuttled to Sonora, in northern Mexico, there was no water for washing or drinking, electricity or "sanitation". Borlaug slept in the loft of a shed that he shared with vermin, insects and snakes. In Chapingo, he cooked in the open while only occasionally travelling to his hotel in Mexico City for a meal and bath. Meals at both research stations consisted of canned food cooked over an open fire. In Sonora, he was often sick from the water, though boiled (Vietmeyer 2011a). Travel to the research sites was arduous with Borlaug carrying his own provisions. Once there, the little available equipment was basic. In Sonora, Borlaug initially used hoes to manually sow the fields and, in Chapingo, he and his colleagues initially pulled the ploughs themselves (Mann 2018).

While often working alone or with a handful of US colleagues, Borlaug had contact with neighboring farmers, but communication was difficult. His command of Spanish was poor and save for a few friendly if not curious farmer neighbors, Borlaug faced skepticism as to his ventures. Only a handful showed up for his first demonstration day in Sonora and no one accepted the free seed that he offered. Despite the adverse conditions and immense challenges, Borlaug persevered.

Being in the field was essential to wheat breeding for Borlaug. His now often-cited mantra of needing to "listen to the wheat", entailed getting your hands dirty. His almost missionary zeal about the need to be hands-on can be seen in the repeated accounts (Hesser 2009; Mann 2018; Vietmeyer 2011a) of his initial exposure to Mexican agricultural scientists in Chapingo, where Borlaug first attempted wheat breeding in Mexico. The account describes reluctance of his first Mexican colleagues to undertake physical work. Pepe Rodríguez, one of the researchers, apparently explained that they were not doing physical research work because "we don't do these things …. You should draw up the plans and take them to the foreman and let the peones do the work." Borlaug retorted "If we don't do things ourselves…how can we advise anyone? How can we ever be certain of our ground?" (Vietmeyer 2011b, p. 93).

For Borlaug, getting your hands dirty was indicative of one's capacity for not only intellectual work as a researcher but also a wider commitment to and sacrifice needed in the pursuit of science. With this came a dress code: according to a number of accounts, the initially reluctant aforementioned Mexican researchers eventually traded in their "suits and shiny shoes" for Borlaug's kit of "khaki pants, lace-up boots, and sweat-stained baseball caps". This became the "project uniform" (Mann 2018) and later dress code for Borlaug adherents and a point of pride that they had caught the "Borlaug bug" (Vietmeyer 2011b, p. 94).

Borlaug research colleagues' clothing choice represented everything Borlaug was not: dressing in suits and shiny shoes was emblematic of their social status and reluctance to take on manual work (Vietmeyer 2011b). In contrast, getting your hands dirty is what constituted a wheat breeder. For example, Borlaug arrived at a news conference in Mexico City when his Nobel prize was first announced wearing the "uniform": his work clothes, dusty shoes along with dirty hands. As he explained later, Borlaug "wanted to show the TV men what makes an agricultural scientist — dirty hands". He later washed them (Perkins and Longden 2009).

The senses of "hard work" and "getting your hands dirty" were manifestations of Borlaug's steadfast belief that "you had to listen to the wheat" and "demonstrating by our own field results what could be done" (Hesser 2009, p. 43). By intimately observing plant growth, researchers can understand them best. For example, for the first trial in Chapingo in April 1945, Borlaug and his two colleagues planted 110,000 plants by hand including 8600 varieties covering over five miles of rows of plants. They were breeding wheat that was resistant to stem rust, so they continuously inspected each plant for stem rust, uprooting any diseased plants. Even with four assistants, inspecting all the plants took at least two weeks, after which the round of inspection would be repeated (Mann 2018). The thousands of crosses were repeated in the winter at irrigated plots in Sonora as part of the shuttle breeding program (Zeyen et al. 2009). During the years of the Office of Special Studies trials, Borlaug and his research team studied about 40,000 varieties and lines, many of which were planted in several locations (Hesser 2009, p. 59). The researchers were looking for a needle in the haystack that required meticulous selection for, as Borlaug instructed, "kernels may be gold nuggets .... Find them!" (Hesser 2009, p. 60). Together—getting your hands dirty and meticulous manual sorting of great numbers of variables—comprised both "fieldwork" and where the real action occurred, and where Borlaug was most at home (Rajaram 2011).

Borlaug's single-mindedness and determination, as exemplified by his approach to fieldwork, is a hallmark to his success as a wheat breeder, and it served him well. As Mann (2018) argues, at the time when he was hired to join the Rockefeller Foundation team in Mexico, he lacked subject expertise and professional reputation. (Borlaug was a plant pathologist by training, not a plant breeder). Borlaug landed the job in part because prior candidates had declined it (Mann 2018). Once in Mexico, he was likely propelled by his capacity to "not be defeated by difficulty" and his "missionary zeal" (Hesser 2009, p. 34).

Throughout his career Borlaug faced resistance to his ideas. For example, his idea of shuttle breeding bucked all conventions and principles of agronomy. When the Rockefeller Foundation first rejected his idea of planting in what was considered out of season and context, he proceeded with wheat breeding in the Yaqui Valley in Sonora on his "own time" in addition to his "day job" and at his own expense (Mann 2018, p. 135). When his colleague, Ed Wellhausen, tried to discourage him given the infeasibility of the idea, Borlaug responded, "Don't try to discourage me, Ed. I know how much work is involved. Don't tell me what can't be done. Tell me what needs to be done—and let me do it .... To hell with the extra work and strain. It's got to be done, and I believe I can do it" (Hesser 2009, p. 50).

Borlaug had a capacity for hard work (Rajaram 2011) and enduring pain, which became a code of honor (Hesser 2009) where resistance meant to continue working and working harder. Borlaug recalls: "Oh, there was an abundance of criticism. I used to tell our team, 'Just don't listen. Keep working'. We could prove those doomsayers wrong" (Borlaug 2008). Part of this is a certain gung-ho attitude as part of the determination to affect change: "You had to have the guts — if you'd permit me to use that gutter term — to make a decision and say, 'We're going to go for it'" (Borlaug 2008).

Complementing drive, determination and a penchant for hard work, Borlaug also exhibited a certain amount of self-doubt. At his low points during his first few years in Mexico, after facing adversity and challenging conditions, he was certain he had made a "dreadful mistake in resigning from [his] former position" (Hesser 2009, p. 42). He confided to his wife, Margaret Borlaug, that he was unsure of what he was doing. In the face of such uncertainty, Borlaug employed determination that was accompanied by a leap of faith. Hesser describes Borlaug's first trials in Chapingo, where he apparently sowed the small sample of seeds and "looked up toward the heavens, crossed his fingers, and said to himself, "It's in your hands now, Lord" (Hesser 2009, p. 49). At his core was a venturesome spirit (Hesser 2009), courage and ultimately self-confidence (Ortiz and Mowbray 2007).

Within the paradigm of heroic individualism, determination and fortitude make for qualities of leadership. They are generally admirable traits yet nevertheless they are gendered: such leadership traits are accepted, even expected, from a man, yet likely unaccepted as innate characteristics if a woman demonstrated these same qualities, particularly in the late 1940s and the 1950s (see Black et al. 2019). Similarly, determination and willingness to work hard, particularly in the face of adversity, allow for discovery and innovation and are certainly not gender-specific capacities. They are gendered, however, in that being able to wholly dedicate your time to a single endeavor is available to those free of competing responsibilities, such taking care of children and undertaking other social reproduction roles.

During Borlaug's era, it was men who had the privilege of being released from reproductive work.[5] Margaret's labor sustained her husband's professional career and accomplishments through her full-time role as family care-giver. While she generally is portrayed in biographies of Borlaug in a secondary role, Hesser (2009) and Mann (2018) substantively discuss her role in Borlaug's life. Mann, in particular, notes her generally preceived absence in his account of Borlaug when he writes "Left unnoted in this description: Margaret would be left to raise their daughter by herself for half the year" (Mann 2018, p. 134). What Mann refers to is what feminist economists analyze as the invisible social reproduction work that sustains the life and the work of others. Such obscurity is illustrated, for example, when Borlaug expressed concern to Margaret Borlaug about the impoverished conditions in Mexico and writes "Can you imagine a poor Mexican guy struggling to feed his family? I don't know what we can do to help these people, but we've got to do something" (Hesser 2009, p. 39). Absent is a consideration of the social reproductive work undertaken by the Mexican guy's wife, who presumably was as or even more struggling to care for the family.

While Borlaug's biographers note Margaret's and other family members' personal "sacrifice" for the sake of Borlaug's professional endeavors, a complete reckoning of her vital role sustaining his work is missing. Ditz (2004, p. 17) argues that

---

[5] For example, when Margaret Borlaug is featured in biographies on Borlaug, it is more about *his* absence from their family life. For example, Phillips recalls how Margaret Borlaug quipped that while they had been married for 60 years, her husband has been home for four. He claims this was "a fairly accurate estimate of the time Borlaug had spent at home" (Phillips 2013, p. 10) given that Borlaug split his year between the Sonora experiment station and Chapingo, near Mexico City where the Borlaug family lived. While Borlaug was an absent parent, he prioritized major family events and coached his son's baseball club during their earlier years in Mexico (Zeyen et al. 2009).

this social relation of gender is critical in pinpointing "what is gendered about the characters and conduct of men … which [historians of masculinity] too often pay only lip service". He suggests that "conceptualizing the gender order in terms of men's access to women is an excellent starting point in a wide variety of historical settings" (Ditz 2004, p. 11).

Borlaug's seminal work took place in the post-war era of the US, and it would be fool-hardy to assess this period as a basis for analyzing subsequent developments in breeding practices thereafter or to retrospectively pass normative judgment of the main gender division of labor at that time. Much has changed in gender relations and roles since then. However, both Borlaug and his influence on the next generation of breeders and the professionalization of wheat researchers extends beyond this period, particularly in the imprinting of what it means to be a wheat researcher and subsequent delimiting of who can be a wheat researcher, in part because of their gender and relative positioning within a gender order. This is explored in the next section.

5.3.2   Reproduction of Borlaug's Work Ethos

Complimentary to Borlaug's unbridled faith in technology and the capacity of "mankind" to overcome adversity, he also was a committed adherent to the notion of educating farmers as well as agricultural researchers. As colleagues Ortiz and Mowbray attest, the "active promotion of scientists in agricultural extension and in intervening in policy formulation … became the hallmarks of Borlaug's work" (2007, p. 17). This section critically examines principles for training that, on the one hand, aimed to produce the next generation of wheat researchers, while reproducing hegemonic notions of masculinity, thereby limiting their effect.

For both Borlaug and the Rockefeller Foundation, training an "army" of young Mexican researchers was key to the country's food security. Foundation staff aimed to work themselves out of jobs, with trainees eventually assuming "primary responsibility for the well-being of agriculture in Mexico" (Borlaug 2008). Borlaug's initial cohort started with the so-called "bird boys": three young boys, hired from the village of a foreman, whose job was to scare away birds from eating the wheat grain. One in particular, Reyes Vega, would help realize Borlaug's ambition. Vega became a successful wheat breeder, leading innovations in pollination that would save the researchers "man-years of effort" (Vietmeyer 2011b, p. 126). Borlaug would later initiate "a dozen of the largely unlettered bird boys into the arts of managing wheat research and of cross pollinations" (Vietmeyer 2011b, p. 135) in Sonora.

Later, the Rockefeller Foundation training program consisted of two main components. One was an internship initiative in Chapingo, attached to the research program, which aimed to train a new generation of scientists (Borlaug 1970). A second component was a fellowship and scholarship program that supported Mexican researchers to undertake graduate studies in the US. The first formal cohort was in 1943 when "a few young Latin American agricultural scientists" spent up to a year with staff of the Rockefeller Foundation/Mexican Ministry of Agriculture

Cooperative Agricultural Program on agronomy and crop improvement. Participants worked alongside senior researchers, after which they returned home with "improved lines or populations of germplasm" to use in their own research (Villareal 1994, p. 112).

According to the Rockefeller archives, from that first cohort in 1943 to 1963, some 550 interns, all men, participated in the agricultural research and training program; 200 received a Master of Science degree and about 30 PhDs with Rockefeller scholarships. Focusing on the most promising scientists, these now trained researchers were to staff the new National Institute of Agricultural Research in 1961 (Hesser 2009). For E.C. Stakman, the University of Minnesota plant pathologist who led the initial Rockefeller Foundation team in Mexico, "The development of a competent corps of Mexican agricultural scientists and scholars was the most valuable permanent contribution of the revolution in agriculture" (Hesser 2009, p. 64).

From the earliest days in Mexico, Borlaug (1970) worked from the premise that there were no trained agricultural scientists in Mexico. Of course, there were trained Mexican agricultural researchers. The first agricultural college in Mexico was established as early as 1854 in Chapingo, where Borlaug undertook his first experimental plots. Borlaug was referring to Mexican agricultural researchers not being trained in his way. From his earliest days, Borlaug "gave" his Mexican colleagues "the opportunity and responsibility to learn, to become proficient in the secrets of plant breeding – crossing and selection – the critical steps that most plant breeders kept to themselves" (Hesser 2009, p. 43). The idea was for Mexican counterparts to work along Borlaug with the emphasis on fieldwork, an agenda that harks back to Borlaug's initial experience with Mexican researchers and their reluctance to forgo their middle-class clothing and don Borlaug's uniform. It was this philosophy—that "the crosses and selections must be done on site, and one should observe the unique characteristics of each variety *throughout its growth cycle*" (Rajaram 2011, p. 26, my emphasis)—that understandably underscored the program and its design.

This approach to wheat research became even more widespread when the internship training of the initial Rockefeller Foundation program became institutionalized as the international training program of CIMMYT, particularly its International Wheat Training Program (IWP) (Rajaram 2011; Villareal 1994). "Borlaug concluded that the same program of training that had helped so many young scientists from Mexico and other Latin American countries to assume leadership in their countries' agriculture could be used to train and motivate young scientists from the Middle East and South Asia" (Hesser 2009, p. 68). With the support from the Rockefeller Foundation and Ford Foundation, the IWP in particular and the wheat research program more generally served as a "template" for the 1960 establishment of the International Rice Research Institute in the Philippines, as well as CIMMYT's training discussed next.

While training future wheat breeders in ways that Borlaug had pioneered, such an approach also reproduced his work ethic. The belief in hands-on fieldwork and practice and eschewing a sole focus on "theory" informed the first formal IWP in Mexico in 1958 (Rajaram 2011), which is the best known of the CIMMYT training offerings among wheat researchers. For example, during the eight-month program, trainees learned about shuttle breeding, moving between Sonora, where they worked February to April, and then to Toluca and Chapingo from May to October. Working 12 hours a day, they were occupied with fieldwork as part of their basic wheat training (Hesser 2009, p. 69).

Other early generation CIMMYT trainings also focused on instilling the value of hands-on research.[6] Villareal (1994)[7] describes how since 1985, 67% of course time is spent in the field and lab practicums, which he describes as important, as most participants had sufficient theoretical knowledge but little practical experience. CIMMYT's comparative advantage over other universities and other institutions was to provide such experience.

Apparently, young scientists working with Norman Borlaug found the training demanding but rewarding and described it as "simultaneously being in the Peace Corps and in a Marine Corps boot camp" (University of Minnesota n.d.-b). What young scientists learned was the urgency of their mission to help feed the world. On a practical level they learned that wheat waits for "no man or woman. When wheat flowers are ready for cross breeding you work from sun up to sun down, because the window of opportunity rapidly closes" (University of Minnesota n.d.-b).

The boot-camp reference seems apt given Swanson's 1975 description of the program and its rationale.[8] To describe the approach, he uses the metaphor of plant breeding where wheat plants that are grown "under a variety of different conditions, both favorable and unfavorable growth environments, [will] respond differently to those conditions" (Swanson 1975, p. 87). A plant breeder needs to observe carefully to select the genetic lines with the greatest potential. CIMMYT's "somewhat controversial" (Swanson 1975, p. 87) approach to training was also a way of selecting the researchers themselves. Trainees are intentionally required to undertake "hard, backbreaking work, wading through muddy plots, many times in the rain" (Swanson 1975, p. 87). Swanson acknowledges that after half a day, "there is no additional technical training value to be accomplished", but trainees are still required to undertake the work for about two weeks. The aim is to learn about the trainees and their "real" ability and attitudes towards the work of breeding. CIMMYT was particularly interested in identifying those trainees willing to do the hard work and who "'identify with' or have 'internalized' *positive* attitudes toward this type of research" (Kelman 1958 cited by Swanson 1975, p. 87, my emphasis).

For CIMMYT, the wheat training program was an "early generation selection tool for identifying potential, hardworking research scientists that are oriented toward practical, problem oriented-field research" (Swanson 1975, p. 87) where

---

[6] To develop the trainees' confidence in their knowledge and skills, their understanding of the importance of working hands-on in the field or in the laboratory, their ability to work in teams, and their awareness of the value of multidisciplinary research (Cooksy and Arellano 2006, p. 12)

[7] R. L. Villareal led the wheat training program at CIMMYT for 18 years.

[8] Burt Swanson was Professor Emeritus at University of Illinois at Urbana-Champaign. He served as CIMMYT's first training officer and worked closely with Borlaug (https://www.linkedin.com/in/burt-swanson-a7752114/). He died in 2020.

Borlaug's "educational hands-on, long days in the field, [sic] philosophy was the backbone of the international training programs at CIMMYT and eventually at all international crop centers" (Zeyen et al. 2009). Practically, this meant that trainees would later be identified for fellowships and additional educational opportunities with the hope that they would assume key research and leadership roles at wheat improvement programs in their home countries.

By all accounts, this aim of influencing wheat research globally was achieved. Later to be known as "wheat apostles"[9] (University of Minnesota n.d.-a), CIMMYT trainees and followers of Borlaug number in the thousands (Zeyen et al. 2009). According to Hesser (2009), total CIMMYT alumni include some four thousand researchers from 120 countries, some of whom became "pillars of wheat research and development programs in 80 developing countries" (NAST 2015, p. 26). Both peers and trainees alike have gone on to influence global wheat breeding, and plant breeding more generally.

In this way Borlaug's legacy goes beyond the technical innovations and extends to the global impact of his professional development approach, which Zeyen et al. (2009) claim "literally changed [the] world in developing and underdeveloped countries". For example, as Hesser (2009, p. 20) states, Borlaug's "fanatical devotion to wheat paid big dividends. Many of his young Mexican associates caught the 'wheat fever' from him, and together they carried the wheat revolution to a successful conclusion". He goes on to claim that Borlaug's emphasis on field-based research became a "model for scientific training in the international (AR4D) centers" (Hesser 2009, p. 124).

At CIMMYT in particular, Borlaug's influence is long-lived. In their 2007 dedication to Borlaug, Ortiz and Mowbray[10] account for CIMMYT's success to the Borlaug's hallmarks, which "still constitute the modus operandi of CIMMYT and reflect his work ethic" (Ortiz and Mowbray 2007, pp. 23–24). For his former colleagues, Borlaug left enduring technical approaches, particularly shuttle breeding, that have been globally disseminated by CIMMYT through their training and interactions with agricultural leaders and researchers. This has motivated fellow scientists from AR4D organizations "to leave their desks and work in the field, where they are able to identify and tackle more effectively the real problems farmers face" (Ortiz and Mowbray 2007, p. 24).

The next section explores the gendered implications of the army of wheat researchers initially envisioned by Borlaug (2008) and later his "apostles".

---

[9] Hesser (2009: 44) also uses this term.

[10] Rodomiro Ortiz is a geneticist and plant breeder and held several senior positions at CIMMYT as well as other international AR4D centers.

5.3.3 Gendered Implications

That Borlaug had an influence on the professionalization of a generation of wheat researchers is not being contested nor is his impact of their work on wheat agronomy and their impact on food security. The question remains, however, who were these "apostles" and what were they preaching? In arguing for a Gendered Archeology of Organizations, Goetz (1997, p. 19) refers to "privileged groups, included and excluded groups, superiors and subordinates" and the need to identify them "in determining which group's interests are served by a particular institution". This section examines the gendered legacy of Borlaug's early work.

Gendered Participants In all accounts of Norman Borlaug's early career, men dominate the array of actors in his life. For example, George Harrar, who was to become Borlaug's boss in Mexico, was the Rockefeller project director. The initial team from 1948 included a three-man agricultural advisory committee: Stakman, Bradfield, and Mangelsdorf (Hesser 2009). The original team Borlaug brought together in Chapingo and Sonora were all men, except Angela Meléndez and Marta Zenteno who worked as research assistants for the initial fieldwork in 1945. They, however, were not allowed to stay overnight in the fields (Mann 2018). Two other exceptions include Dorothy Parker, the librarian, and Evangelina Villegas, biochemist in the Office of Special Study and the 2000 recipient of the World Food Prize.

Men not only dominated Borlaug's peer group but also CIMMYT trainees. CIMMYT's database lists 1858 participants during his tenure at CIMMYT from 1966 to 1979, coming from about 100 different countries, in various formal training programs, including the basic and advanced wheat improvement course (with various specializations such as pathology, breeding and bio-technology). Of these, 242 (13%) were women, with the first women attending in 1970, some 20 years after the first cohort of trainees.[11] Most women participants came from China, Pakistan, India, Mexico, Tunisia and Turkey and to a lesser extent Egypt, Kenya, and Kazakhstan.

That men dominated the sector during this period and after, whether at CIMMYT or elsewhere,[12] is indicative of a particular time. Still, the issue is to not only understand the basis for this gender gap, as well as its reproduction, but also to understand the fixing of this trajectory and its gendered implications, discussed next.

---

[11] Analysis based on CIMMYT database accessed in 2022. The proportion of women in CIMMYT training increased over the years. Additionally, CIMMYT introduced the Women in Triticum program, which included 37 participants from 2013 to 2020.

[12] For example, the International Rice Research Institute (IRRI), established in 1960 as the second international agriculture research center after CIMMYT, had its first and only woman PhD researcher, Kwanchai A. Gomez, in 1968. The next was in 1978. Correspondence with Margreet van der Burg, based on the IRRI Staff Listings, 1961–2005.

Gendered Ways of Working In their gender audit of CIMMYT, Merrill-Sands et al. (1999) observed how its early history continued to strongly influence the organization's work culture and values. These included staff referring to the "sacrifice and selfless devotion, of the mission of the organization taking priority over everything else, including family and personal life" (Merrill-Sands et al. 1999, p. 12). They identify four mental models that characterized CIMMYT as an organization, inherited from earlier times[13] including "the ideal CIMMYT worker", which was engrained in CIMMYT's past. Such a worker "was instilled with missionary zeal, willing to sacrifice everything and endure hardship to get the job done...[spend] time in the field" with the assumption that CIMMYT workers "did not have competing responsibilities in private life" (Merrill-Sands et al. 1999, p. 16).

While not diminishing the importance of hands-on research through fieldwork and the importance to understand a specific context firsthand, such acts need to be understood as part of hegemonic masculine practices that reproduce notions of what it is to be an "ideal worker", which in the case of AR4D, translates as a valid researcher. As seen above, "fieldwork" in CIMMYT and its training of wheat researchers became a rite of passage and normalized. Liebrand and Udas (2017) likewise note in their study of water engineers in Nepal how fieldwork became "strongly associated with normal professional performance" (Liebrand and Udas 2017, p. 6) and "is the paradigmatic experience for constructing and reconfirming male and engineering identities" (Liebrand and Udas 2017, p. 9) of ingenuity and masculinity.

Undertaking arduous fieldwork for wheat breeding involving constant travelling, initially all by men then later dominated by men, are performances of masculinity: beyond its functionality, fieldwork serves to simultaneously re-enact and also re-experience a previously established accepted way of working, such as introduced during and codified by CIMMYT training. The unquestioned repetition of such practice reproduces the legitimacy of what is to be a wheat researcher, based on foundational practices that men modelled, as initially exemplified by Borlaug.

Wheat researchers who undertake annual migrations within Mexico (and often internationally) constitute a *community of practice*. The performance of fieldwork is routinized to such an extent that it is normalized by collective agreement and providing a sense of belonging. This is not to say that wheat agronomists are "cultural dopes" (Lynch 2016), blindly following prescribed working norms. Rather, there is both consent and coercion, by establishing and then following "ways of doing things". This process of consent and coercion—hegemony—is also gendered, producing hegemonic masculinity: initially as only men undertake such acts and later, as part of wheat researchers' training, as a masculinized norm for working. As a community of practice, this is both inclusionary, through a sense of being accepted and belonging, while at the same time exclusionary for those who could not fulfill

---

[13] Merrill-Sands et al. (1999, p. 12) do not name Borlaug but refer to staff reflecting the time of the Green Revolution.

norms of work, as judged by others, such as the training supervisor, or by their own cognizance, as the personal account at the beginning of this chapter suggests.

This is not to say that all researchers practiced the same approach, but rather particular ways were privileged if not actively used as part of a selection process. Such norms contribute, at least at CIMMYT, to behavior that "no one questioned the belief that to do good science, one had to work extremely long hours at far-flung field research sites and to place work ahead of all else. Nor did they question the gendered impact of such assumptions" (Rao et al. 1999, p. 8). There are at least two.

One is the implicit exclusion of women, as illustrated by the introductory quotes. The work patterns and ethos that characterized Borlaug's work are aligned with the CIMMYT "ideal worker": traveling incessantly, working long hours. Men are not the only ones who can hard work, for long hours and travel extensively, but to be able to do so, one must be free of social care responsibilities and have support from others. Within the dominant gender division of labor, and women bearing much of the responsibility for childcare and eldercare, this way of working limits and excludes many of them. Men are the privileged gender to perform hegemonic masculinized ways of working with the tacit support of others; women are faced with difficult and complex decisions and trade-offs, particularly when they do not have the support of other caregivers or they choose not to avail themselves of such support.

Such exclusionary and delineating effects take on particular dimensions in the context of transglobal AR4D as in the case of international agricultural research organizations. For example CIMMYT, based in Mexico, is still the epicenter of wheat and maize research, where participating in professional development at its headquarters and field stations in Mexico is a rite of passage for wheat and maize researchers and a badge of honor. It also has gendered implications for who can take advantage of such opportunities: not all wheat or maize researchers can travel internationally, unencumbered by socio-cultural gender norms. Moreover, with the headquarters based in the Central Standard Time zone, virtual meetings are often in the evenings for colleagues in the Middle East, Africa and Asia. This requires either staying late in the office or working from home; two working conditions that are gendered. For personal security as well as prevailing gender norms and roles, only certain scientists, mostly men, can participate in such meetings. Ultimately, as a form of international collaboration, this influences the professional advancement of those who can be visible to and access senior colleagues based in the headquarters.[14]

---

[14]This paragraph is based mainly on the author's observations having researched CIMMYT and worked with staff in Mexico and in its programs in Kenya, Zimbabwe, India, and Nepal. While there is a dearth of gendered analysis of professional development and advancement of AR4D researchers globally, studies mainly from the global north suggest that women researchers face gendered constraints, in part due to reproduction responsibilities and gender norms within their personal and work contexts, which can hinder international mobility and cooperation and thus limit professional advancement (Elsevier 2017; Kwiek and Roszka 2021; Uhly et al. 2017). The shift to more virtual work in some cases overcomes these constraints but, in others, exacerbates the gendered effects of temporal and spatial distances of global work (Villamor et al. 2023).

Another related gendered implication is the impact on men. A hegemonic masculinity analytical frame is also helpful in that the dominant ways of working are just that: dominant. This is not to say that all men follow these work habits, but there is a hierarchy of ways of working for men, and it comes with costs. While there is little research on the social impacts of ways of working in AR4D, Bryant (2021) and Brandth (2021) document the adverse impact on men farmers' mental health from upholding the "idealized character of male embodiment in agriculture, which assumes a high degree of strength and stoicism" (Brandth 2021, p. 388). For Australian farmers, upholding "hegemonic masculinities that render their bodies as strong, stoic, and resilient and pitted against nature" (Bryant 2021, p. 425 citing Alston 2011) are not far removed from Borlaug's ways of working, which he cultivated in trainees. Additionally, others have documented the emotional and psychological impacts on both men as fathers as well as their families, particularly children, with absent fathers.

5.4   Conclusion

In this chapter, I have described how Borlaug faced challenges in the early days of his wheat research with determination, strong will and a penchant for enduring arduous work. Often to prove naysayers wrong, he had to double down. His approaches to wheat breeding included his dedication to fieldwork and shuttle breeding required continuous travel. This hard work was demanding of researchers, all men during that time, who, like Borlaug, had to make sacrifices in their personal lives. Later, the embedding of such a work ethos in wheat breeding were reproduced in the professional development of subsequent wheat researchers, mainly through the Rockefeller Foundation scholarship program and later through international CIMMYT wheat-breeding courses. These professional development opportunities that aimed to train a global army of wheat researchers were, by all accounts, successful in that trainees, mostly men, went on to become agronomists. Some became leading researchers in their own countries, and beyond.

To be sure, Borlaug was motivated by compassion for and a commitment to eradicating poverty and suffering as he himself witnessed in the US during the Great Depression and later when he first started working in Mexico and South Asia. The purpose of this chapter is not to refute Borlaug's achievements but extend his legacy by also unearthing previously unaccounted for contributions to better understand how AR4D has historically been constituted, and how that, in turn, has made change and gender integration particularly difficult.

I have attempted to account for the imprinting of specific ways of working, and working with a particular group of men that gets reproduced through professional development and advancement. This process of reproduction has gendered dimensions in that it is generally men who are free of social reproduction roles who can avail themselves of such demanding opportunities and advance their careers as wheat breeders. Conversely, those who do undertake such reproductive roles, mainly women, are more challenged in succeeding as agronomists within the prevailing work paradigm. I have also suggested that upholding such masculinized norms of work can have adverse health and personal well-being implications for men and others who pursue such norms.

The mobilization of the ideal wheat researcher, through the deployment of discursive tropes of what constitutes becoming a researcher through repetitive, stylized acts such as fieldwork, are concerned with gender performativity. The "project uniform", nomadic lifestyles and other tangible if not intangible representations of having performed fieldwork produce, through their repetition, a certain homogeneity and hegemonic masculinized notion of what constitutes a wheat researcher if not wheat research. These norms are reproduced through communities of practice of training, the establishment of initial and on-going professional networks of wheat breeders as well as the telling and re-telling of Borlaug hagiography and related stories that breeders share with each other, such as the numerous dedications to Borlaug's legacy. In addition to bearing witness to such a revered man, they also serve as a basis for, on the one hand, shared identification and belonging while, on the other, exclusion.

In undertaking this gendered archeology (Goetz 1997), I do not attempt to attribute gender dimensions of contemporary wheat agronomy or AR4D to the past. That is not necessary, given the dramatically different social and AR4D contexts of today and the many influences on AR4D beyond Borlaug. Rather I have analyzed the past to re-make the future. However, the privileged concepts, values, history and practices highlighted above can also be understood as "gender-specific metaphors [that] must be integrated into the analysis of relations of domination and subordination" (The Editorial Collective 1989, p. 6). They are "powerful motifs of hegemonic masculinity" (Vera-Gajardo 2021, p. 10) commonly found in science and technology that serve as the basis for the continued institutionalized dominance of men over women (Page et al. 2009) based on "assumptions that reflect the values and life situations of men and of idealized masculinity" (Merrill-Sands et al. 1999, p. 11). Such an analysis points to the need for systemic change that includes, but also goes beyond, efforts to "fix women", such as training them in leadership, or that support adaptation to what is otherwise a gender-biased system, such as policies to achieve "work-life balance".

Not included in this historic reckoning, though potentially illuminating, is an extended gender analysis of Borlaug's innovations as the basis for early transglobal positivist and modernist-oriented AR4D as well as an analysis of the critical and related roles of philanthropic capitalism, e.g., the Rockefeller Foundation and its role in establishing global AR4D. As Hamilton et al. (2017, p. 613, my emphasis) contend, "science is constitutive of colonialism … (that) … must be understood through the *histories* of science".

References

Åsberg C, Lum J (2010) Feminist technoscience studies. Eur J Women's Stud 17(4):299–305

Baranski M (2022) The globalization of wheat. A critical history of the green revolution. University of Pittsburg Press, Pittsburg

Barbarcheck M (2021) Gender and agricultural extension. In: Sachs C, Jensen L, Castellanos P, Sexsmith K (eds) Routledge handbook of gender and agriculture. Routledge, New York, pp 225–238

Black S, Estrada C, de la Fuente MC, Orozco A, Trabazo A, de la Vega S, Gutsche RE (2019) Nobody really wants to be called bossy or domineering. Journal Pract 13(1):35–51

Borlaug N (1970) The green revolution, peace, and humanity

Borlaug N (2008) The father of the green revolution. https://achievement.org/achiever/norman-e-borlaug/#interview

Brandth B (2021) Embodied work in agriculture. In: Sachs C, Jensen L, Castellanos P, Sexsmith K (eds) Routledge handbook of gender and agriculture. Routledge, New York, pp 383–393

Bryant L (2021) Farming, gender, and mental health. In: Sachs C, Jensen L, Castellanos P, Sexsmith K (eds) Routledge handbook of gender and agriculture. Routledge, New York, pp 421–434

Butler J (1988) Performative acts and gender constitution: an essay in phenomenology and feminist theory. Theatr J 40(4):519–531

Connell RW (1993) Book reviews: meanings for manhood: constructions of masculinity in Victorian America by Mark C. Carnes, Clyde Griffen; recreating sexual politics: men, feminism and politics Victor J. Seidler Signs 19(1):280–285

Connell RW, Messerschmidt JW (2005) Hegemonic masculinity: rethinking the concept. Gend Soc 19(6):829–859

Cooksy LJ, Arellano E (2006) CIMMYT's formal training activities: perceptions of impact from former trainees, NARS research leaders, and CIMMYT scientists. CIMMYT, Mexico DF. https://pdf.usaid.gov/pdf_docs/Pnado623.pdf

Crewe E, Harrison E (1998) Whose development? An ethnography of aid. Zed Books, London

Ditz TL (2004) The new men's history and the peculiar absence of gendered power: some remedies from early American gender history. Gend Hist 16(1):1–35

Elsevier (2017) Gender in the global research landscape. Elsevier, Amsterdam

Farhall K, Rickards L (2020) The "gender agenda" in agriculture for development and its (lack of) alignment with feminist scholarship. Front Sustain Food Syst 5:1–15. https://doi.org/10.3389/fsufs.2021.573424

Faulkner W (2009) Doing gender in engineering workplace cultures. II. Gender in/authenticity and the in/visibility paradox. Eng Stud 1(3):169–189

Goetz A-M (1997) Introduction. In: Goetz A-M (ed) Getting institutions right for women in development. Zed Books, London

Hamilton JA, Subramaniam B, Willey A (2017) What Indians and Indians can teach us about colonization: feminist science and technology studies, epistemological imperialism, and the politics of difference. Fem Stud 43(3):612–623

Haraway D (1988) Situated knowledges: the science question in feminism and the privilege of partial perspective. Fem Stud 14(3):575–599

Hesser L (2009) The man who fed the world, 2nd edn. Righter's Mill Press, Princeton

Kelman HC (1958) Compliance, identification and internalization, three processes of attitudinal change. J Confl Resolut 2:51–60

Kwiek M, Roszka W (2021) Gender disparities in international research collaboration: a study of 25,000 university professors. J Econ Surv 35(5):1344–1380

Laveaga GS (2021) Beyond Borlaug's shadow: Octavio Paz, Indian farmers, and the challenge of narrating the Green Revolution. Agric Hist 95(4):576–608

Liebrand J (2014) Masculinities among irrigation engineers and water professionals in Nepal. PhD thesis. University of Utrecht

Liebrand J, Udas PB (2017) Becoming an engineer or a lady engineer: exploring professional performance and masculinity in Nepal's Department of Irrigation. Eng Stud 9(2):120–139

Lynch M (2016) 'Cultural dopes'. In The Blackwell encyclopedia of sociology, pp 1–2. https://doi.org/10.1002/9781405165518.wbeos0712

Mann CC (2018) The wizard and the prophet: two remarkable scientists and their dueling visions to shape tomorrow's world. Vintage Books, New York

Merrill-Sands D, Fletcher J, Acosta A (1999) Engendering organizational change: a case study of strengthening gender equity and organizational effectiveness in an international agricultural research institute. In: Rao A, Stuart R, Kelleher D (eds) Gender at work: organizational change for equality. Kumarian Press, West Hartford, pp 77–128

Mukhopadhyay M, Prügl E (2019) Performative technologies: agricultural research for development and gender. Int Fem J Polit 21(5):702–723

NAST (2015) NAST 37th annual scientific meeting "the challenges of non-communicable diseases (NCDs): responding through multisectoral action". National Academy of Science and Technology, Manila. https://www.nast.dost.gov.ph/images/pdf%20files/Publications/ASM/NAST%202015%2037th%20Annual%20Scientific%20Meeting.pdf

Ortiz R, Mowbray D (2007) Dedication: Norman E. Borlaug the humanitarian plant scientist who changed the world'. Plant Breed Rev 28:1–37

Page MC, Bailey LE, Van Delinder J 2009 The blue blazer club: masculine hegemony in science, technology, engineering, and math fields. Forum on Public Policy Online 2. https://eric.ed.gov/?id=EJ870103

Perkins J, Longden T (2009) Nobel laureate Iowa native Norman Borlaug world food prize. The Des Moines Register https://eu.desmoinesregister.com/story/news/2017/09/12/nobel-laureate-iowa-native-norman-borlaug-world-food-prize/658320001/

Phillips RL (2013) Norman E. Borlaug: a biographical memoir. National Academy of Sciences. Washington, DC

Pyburn R, van Eerdewijk A (eds) (2021) Advancing gender equality through agricultural and environmental research: past, present and future. IFPRI, Washington, DC

Rajaram S (2011) Norman Borlaug: the man I worked with and knew. Annu Rev Phytopathol 49:17–30

Rao A, Stuart R, Kelleher D (1999) Roots of gender inequality in organizations'. In: Rao A, Stuart R, Kelleher D (eds) Gender at work. Organizational change for equality. Kumarian Press, West Hartford

Rap E, Oré MT (2017) Engineering masculinities: how higher education genders the water profession in Peru. Eng Stud 9(2):95–119

Sachs CE, Jensen L, Castellanos P, Sexsmith K (eds) (2021) Routledge handbook of gender and agriculture. Routledge, New York

Sefyrin J, Elovaara P, Mörtberg C (2018) Feminist technoscience as a resource for working with science practices, a critical approach, and gender equality in Swedish higher IT educations. 13th IFIP international conference on human choice and computers (HCC13) AICT-537, pp 221–231. https://hal.inria.fr/hal-02001957

Sumberg J, Keeney D, Dempsey B (2012) Public agronomy: Norman Borlaug as "brand hero" for the Green Revolution. J Dev Stud 48(11):1587–1600

Swanson BE (1975) Evaluation of the CIMMYT wheat training program. J Agron Educ 4(1):85–89

The Editorial Collective (1989) Why gender and history? Gend Hist 1(1):1–6

Tickamyer AR, Sexsmith K (2019) How to do gender research? Feminist perspectives on gender research in agriculture. In: Gender, agriculture and Agrarian transformations. Routledge, New York, pp 57–72

Tomar SMS (2009) Norman E. Borlaug, the father of Green Revolution who saved millions from starvation (25 March 1914–12 September 2009) obituary. Indian J Genet Plant Breed 69(4):403–403

Tuns P (2009) Borlaug proved Malthus wrong. The Interim https://theinterim.com/issues/society-culture/borlaug-proved-malthus-wrong/

Uhly KM, Visser LM, Zippel KS (2017) Gendered patterns in international research collaborations in academia. Stud High Educ 42(4):760–782

University of Minnesota (n.d.-a) Taking the Mexican "miracle wheat" to the world's farmers: 1961–1969. https://borlaug.cfans.umn.edu/borlaug/1961-1969

University of Minnesota (n.d.-b) The significance of Borlaug. https://borlaug.cfans.umn.edu/about-borlaug/significance

van Ginkel M, Lillemo M, Trethowan RM (2002) Guide to bread wheat breeding at CIMMYT. Wheat Special Report No. 5. CIMMYT, Mexico DF. https://www.researchgate.net/publication/275148186_Guide_to_Bread_Wheat_Breeding_at_CIMMYT_Wheat_Special_Report_No_5

Vera-Gajardo A (2021) Belonging and masculinities: proposal of a conceptual framework to study the reasons behind the gender gap in engineering. Sustain For 13(20):11157

Vietmeyer N (2011a) Borlaug. The wheat whisperer, 1944–1959. Bracing Books, Lorton

Vietmeyer N (2011b) Our daily bread: the essential Norman Borlaug. Bracing Books, Lorton

Villamor IN, Hill S, Kossek EE, Foley KO (2023) Virtuality at work: a doubled-edged sword for women's career equality? Acad Manag Ann 17(1):113–140

Villareal R (1994) Wheat improvement training at CIMMYT. In wheat breeding at CIMMYT: commemorating 50 years of research in Mexico for global wheat improvement, CIMMYT. https://repository.cimmyt.org/bitstream/handle/10883/1203/56508.pdf?sequence=1&isAllowed=y

Whitehead SM (2002) Men and masculinities: key themes and new directions. Polity, Cambridge

Zeyen R, Ishimaru C, Dickman M, Richardson M (2009) Norman Borlaug: plant pathologist/humanitarian. APS Feature. https://www.apsnet.org/edcenter/apsnetfeatures/Pages/NormanBorlaug.aspx

Chapter 6
Navigating the Patriarchal Politics of Institutions: Positioning Women and Gender Equality at the Center of Agricultural Development Institutions

Jemimah Njuki (ID), **Susan Kaaria, Ednah Kangogo, Kenneth Macharia, Hazel Malapit, Michèle Mboo-Tchouawou, and Sonja Tanaka**

Abstract While the lens of patriarchy has been used to understand the gendered power dynamics at different levels of society, research and development organizations have not been subjected to the same analysis. This chapter explores the gendered nature of organizations, using four key elements of patriarchal organizations (i) a gendered division of labor and tasks, (ii) gendered hierarchies, (ii) rigid rules governing performance, and (iv) privileging male dominated content and processes. We discuss the implications of these characteristics on women's engagement in research, as well as on the positioning of gender research in organizations. Using two case studies, we discuss the key ways in which organizations can be more transformative. Although different organizations are implementing strategies to address gender inequalities in organizations, including through workplace policies and mentoring and leadership programs, we conclude that these initiatives fall short of addressing the structures within organizations and that perpetuate gender-based discrimination.

J. Njuki (✉)
United Nations Entity for Gender Equality and the Empowerment of Women and Girls, UN Women, New York, NY, USA
e-mail: jemimah.njuki@unwomen.org

S. Kaaria · E. Kangogo · K. Macharia · M. Mboo-Tchouawou
African Women in Agriculture Research and Development, Nairobi, Kenya

H. Malapit
International Food Policy Research Institute, Washington, DC, USA

S. Tanaka
Global Health 5050/University College London, London, UK

© The Author(s) 2025   95
J. Njuki et al. (eds.), *Gender, Power and Politics in Agriculture*,
https://doi.org/10.1007/978-3-031-60986-2_6

6.1   Introduction

While there have been many studies on patriarchy and power and how these have shaped society, the concept of patriarchy has not often been applied to organizations. Indeed, the issue of patriarchy in research has been approached as the study of the way in which society is organized and as a topic of research. It has rarely been applied to organizations and specifically the social and organizational context of how research is performed and the power dynamics that govern the organizations carrying out the research and development endeavor. As Witz (1990) in the book "Professions and Patriarchy" says these two words—patriarchy and organizations—are rarely put together.

The concept of patriarchy has been used to refer to a societal-wide system of social relations of male dominance (including those in the family or household). An important concept of the analysis of the patriarchy is that patterns of male dominance in modern society do not rest solely on the unequal distribution of power in the family but extend to other aspects of society including organizations (Walby 1998). And despite the feminist recognition that hierarchical organizations are an important location of male dominance, most analysis of organizations assume gender-neutral organizational structure (Acker 2006). There is an assumption that the patriarchal system is not replicated on organizations—in how they are structured and the power dynamics therein. And yet patriarchy encompasses a highly complex and shifting nature of gender relations, in various sites of social relations such as the family, labor market and state. Acker (2006) further argues that images of men's bodies and masculinity pervade organizational processes, and this has often led to the marginalization of women and contributed to the maintenance of women's segregation in organizations.

Research and development organizations while focusing on the study of gender and class, have only recently started to subject themselves to the same analysis to understand how race, class and gender have interacted in complex ways to produce hierarchies of power and prestige in their work. Witz and Savage (1991), in their paper on gendered organizations have linked the issues of gender and bureaucracy bringing together both organizational and feminist theories to explore how organizations are gendered. A sociological analysis of gender and professions which incorporates a more sophisticated conceptualization of the ways in which gender is itself both socially constructed, and a structuring principle is long overdue.

This chapter argues that organizational structure is not gender neutral; on the contrary, assumptions about gender underlie the organization and practice in organizations including research and development organizations. Notions of patriarchy within research organizations define the basis of the "research endeavor" from questions and notions of who a researcher is, to the organization of research, who gets heard in research and how research is evaluated. We answer the question of what is a patriarchal organization by identifying four characteristics (i) A fixed division of labor or segregation of tasks, (ii) hierarchy of offices, (iii) a set of rules governing performance, and (iv) privileging of male dominated sectors and processes. This chapter uses this characterization and applies it to show how research and development organizations exhibit some forms of these patriarchal characteristics. The paper further describes some of the ways in which these characteristics affect women in research as well as the conduct of gender research.

The chapter provides two case studies, one on the Global Food 5050 accountability report and index showing how such a mechanism can lead to changes in power dynamics within organizations, and the second, the African Women in Agriculture Research and Development (AWARD) and how it is shaping women's leadership and the conduct of gender research in African research organizations. The last section of the paper provides reflections on some of the key elements for addressing power dynamics within organizations drawing on these case studies and draws more broadly on how patriarchal tendencies in organizations are much more complex to address and must start from understating that these structures and power dynamics are and how they are manifested within organizations.

6.2   The Nature of Patriarchal Organizations

Patriarchal organizations are structured around a system of power and authority that is based on gender and reinforces gender-based power imbalances. Gendered hierarchies are reinforced and entrenched through mundane, often textual, organizational processes, such as the division of labor, wages, performance evaluations, and even job descriptions (Acker, 2006). They are characterized by a range of practices and policies that perpetuate gender inequality.

For example, these organizations may have policies that restrict women's opportunities for career advancement, such as limiting their access to training or mentoring opportunities or promoting men to leadership positions over equally or more qualified women. Patriarchal organizations may also perpetuate gender-based discrimination and harassment, which can create a hostile work environment for women and contribute to their underrepresentation in leadership positions. Patriarchal organizations can also reinforce gender stereotypes and norms that limit women's roles and contributions within the organization. For example, women may be expected to take on more administrative or supportive roles, while men are expected to take on more leadership or technical roles. This can limit women's opportunities for professional development and can reinforce gender-based power imbalances.

6.2.1 Division of Labor and Segregation of Tasks

One key feature of patriarchal organizations is occupational segregation by sex which is not unique to the research and development space, nor the agriculture and food systems sectors. As argued by Hartman (1979), and others, the persistence of occupational segregation is explained by the confluence of two systems, patriarchy and capitalism. There are two forms of occupational segregation, vertical segregation describes the clustering of men at the top of occupational hierarchies and of women at the bottom, and horizontal segregation describes that at the same occupational level (that is within occupational classes, or even occupations themselves) men and women have different job tasks.

Looking at how this division of labor and segregation of tasks manifests in the research world we see parallels to this. While the current representation of women in research globally varies by field and country, in general women remain underrepresented in many areas of research. According to data from UNESCO's Institute for Statistics (UNESCO 2015), globally women represent about 30% of researchers. In some fields, such as social sciences and humanities, women's representation is higher, with women accounting for around 45% of researchers. In contrast, in fields such as engineering and computer science, women remain significantly underrepresented, accounting for only around 20% of researchers. These numbers have not changed significantly in the last 8 years. In the agriculture sector this pattern is replicated. Data from African agriculture research organizations shows that only 1 in 5 of agriculture researchers are women, and while this is changing, that change is slow and erratic. Data from the International Food Policy Research Institute's (IFPRI) Agricultural Science and Technology Indicator (ASTI) database indicate that there is recognition of a slight improvement in gender balance in most AR4D institutions. In Kenya, for instance, between 2008 and 2016, the proportion of female researchers in the agriculture sector rose from 25% to 30% and in Nigeria, from 25% in 2008 to 29% in 2014 (Beintema 2017).

Studies have shown that the access to and integration into career networks, distribution of labor in the institution, promotion, and leadership all constrain women's careers, (Hart 2016). Other explanations are the relegation of women to occupations such as nursing and teaching, because women are associated with caring. These explanations with unreconstructed notions of 'women's role' and have no theory of gender relations beyond a basic, taken-for granted 'sex role theory'. And very often, when there is a focus on women's increasing participation in male dominated professions, there is a tendency to focus on the problems women have in adjusting to typically male career patterns, problems which are assumed to be largely generated by the difficulties of reconciling a career with a family (Fogarty, Allen and Walters 1981). In short, the 'dual role' problematic, which focuses on conflicts between family and work roles experienced by women, and which was the dominant focus in studies of women's employment in the 1960s, lingers on in studies of women in 'top jobs'. And yes, women's unpaid care work and family roles are a great contributor to whether women can enter the job market, or what roles they take in the job market, this approach has long been subject to considerable critique (e.g. Beechey 1979) for its neglect of the many other structural factors in the labor markets that constrain women's engagement.

There has also been recent research, focused on the agriculture sector aimed at identifying some of the barriers to women's participation and leadership in the agricultural sector (Mbo'o-Tchouawou et al. 2019; Kaaria et al. 2016). Njuki and Bukachi (2021) describe a range of factors and barriers to gender equality in reference to higher education institutions. These include sociocultural factors, where traditional and cultural norms and practices continue to define the roles of men and women in public life, and what men and women can and cannot do; institutional factors such as environments that perpetuate gender bias, institutional practices, including policies that discriminate and/or exclude certain groups; and economic factors. These studies have shown the existing inequalities between men and women, arising from the patriarchal structure of many African societies and cultural practices and customs assigning certain roles that impose greater constraints on women's leadership capacities. Most of these obstacles are so deeply entrenched in societal norms, customs, and individual mindsets that they require systemic transformation in sociocultural and socioeconomic structures, policies, and practices for change to occur. The absence of women at the highest levels of the scientific hierarchy is an indicator highlights the dysfunction of a system for the evaluation of scientific excellence that has not abolished or weakened the old boy network of co-optation.

These patriarchal tendencies especially within research organizations have contributed to who is recognized as a researcher or a leader. The definition of a researcher can have a gender bias because it often reflects the historical dominance of men in the field of research. Traditionally, the term "researcher" has been associated with men who hold advanced degrees and are employed in academic or research institutions. This definition has historically ignored the contributions of women to research and has reinforced gender stereotypes that associate research with men. Moreover, the definition of a researcher often assumes a certain set of skills and characteristics that may not necessarily apply to everyone. For example, the definition may assume that researchers must be analytical, objective, and competitive, which are often associated with masculine traits. This can often create a hostile work environment for women who may not fit into these stereotypes or may face discrimination based on their gender. It is therefore important to redefine the term "researcher" in more inclusive terms that reflect the diversity of researchers and their contributions to the field. This can include acknowledging the contributions of women and other underrepresented groups to research, as well as recognizing the range of skills and characteristics that are valuable in research, regardless of gender. A much larger implication of these biases within the research system is on who gets funding for research. A review of by Jackson et al. (2022) of science granting councils in Africa showed more awards overall were given to men (51%) compared to women (49%). The review also showed that for research awards specifically, more were awarded to men (62.8%) compared with women researchers (37.2%). Furthermore, awards of a higher monetary value were more likely to be awarded to men compared to women.

6.2.2   Gendered Hierarchies

A second key feature of patriarchal organizations is gendered hierarchies. In a patriarchal organization, men typically hold most leadership positions, and they have
more decision-making power and influence than women. This hierarchical system
assumes that men are more capable and qualified to lead and make important decisions than women, effectively reinforcing gender inequality and discrimination.
Farh and Cheng (2000) have called this paternalistic leadership which they charac-
terize as having three components: authoritarianism, benevolence, and moral leadership. Recent reports in the agriculture sector show the dichotomy of women's
engagement in and leadership in the agrifood sector. The FAO report on status of
women in agri-food systems (FAO 2023) shows that 66% of women in Africa work
in the Agri-food sector, while this proportion is 71% in Southern Asia. At the same
time, the Global Food 5050 report by the International Food Policy Research institute, Global Health 5050 and UN Women shows that only 2% of board chairs of
global food systems organizations are women from low- and middle-income countries (Global Health 5050, IFPRI and UN Women 2022).

Gendered hierarchies are also seen within a vertical segregation perspective,
where there is higher representation of men in organizations that make decisions
within sectors or industries. In the research sector, this is manifested in higher prop-
ositions of men in decision making authority in science granting councils. For
example, Jackson et al. (2022) found that the organizations mandated to disburse
research funds at national and regional level employed more men than women (64%
men and 36% women) and the proportion of women declined as the level of management increased. Most of the science granting councils had very limited funding
programs to eliminate the barriers that women scholars face. This resulted in persistent inequalities in who received funding, the size of the grants they received, and in
the knowledge production, collaboration, and the impact on their country's gender-
related research. Other studies have also shown that women hold fewer positions in
leadership, and this is seen across sectors, and countries. For example, in the US,
while women represented 58.4% of the US workforce in 2022, only held 35% of
senior leadership positions. Only 10.4% of Fortune 500 CEOs were women and less
than 1% of those Fortune 500 CEOs were women of color (Zippia 2023).

6.2.3   Rigid Rules Governing Performance

A third feature of patriarchal organizations is the presence of a set of rigid rules
governing performance, that favor certain groups over others. Ramsay and Parker
(1991) liken these kinds of bureaucratic ways of doing things that is characteristic
of many patriarchal organizations as "rationalized patriarchy". Research and development organizations, like other organizations, are invested in measurement practices. Metrics are crucial to a variety of organizational processes including

standardization, rationalization, governance, recognition and performance management (Weber 2016; Timmermans and Epstein 2010; Scott 1998). But these rules and metric can also be leveraged for the expression of power (Espeland and Stevens 2007). Global audit systems of research that are based on metrics such as citations fail to recognize the multiple and different ways in which men and women contribute to the research and scientific endeavor and the value of different types of research. The system of judgment employed in bibliometrics privileges well-established fields with long-standing publication traditions and clear boundaries. The validity of the science citation index regarding scientific excellence rarely includes sources in multiple languages and covers only a minority of the scientific journals in humanities and the social sciences.

6.2.4 Privileging Male Dominated Content and Process

A fourth feature is the privileging of male dominated content and processes, manifested in what the organization deems as important. In research this is often reflected by the divide between the biophysical and social sciences. The social sciences, including gender research, are often seen as the "soft sciences" and biophysical sciences as "hard sciences". The social sciences have often been seen as a service of the biophysical sciences to explain context in relation to the biophysical sciences (MacMynowski 2007). It is not surprising that the differences in paradigmatic stances of the biophysical sciences and social sciences have engendered status conflicts, and even some negativity, such as through the pejorative use of the terms "hard" and "soft" as described by (Guba 1990; Hedges 1987). Researchers have suggested alternative teams that are status neutral including the use of "high consensus" versus "low consensus" disciplines to better distinguish these differences and avoid such exacerbating language. Set in contrast to the biophysical or natural sciences, the social sciences have often also been portrayed as disunified, in constant conflict, or poorly developed in their theoretical foundations.

Researchers argue that gender mainstreaming, or in this case gender research has been stymied by gendered power differentials and a failure to challenge dominant structures within organizations (Benschop and Verloo 2006; van Eerdewijk 2014). Gender research is often seen as less important or less "serious" than other areas of research, particularly in fields such as science, technology, engineering, and mathematics. This can create a bias against gender research within research organizations and can limit the opportunities for researchers who are interested in pursuing gender-related research questions. The lack of attention to gender research within research organizations has significant implications for the quality and relevance of research. It can limit the diversity of perspectives and approaches that are brought to research questions and can lead to gaps in understanding important social and economic issues related to gender.

Patriarchal organizations create barriers for women to advance their careers and achieve leadership positions but also privilege what research is deemed important.

In a patriarchal organization with hierarchy, men's voices and perspectives are often prioritized over those of women, which can lead to a narrow and limited understanding of the organization's goals and objectives. This can also lead to a lack of diversity in decision-making and contribute to the perpetuation of gender-based power imbalances.

6.3   From Patriarchal to Transformative Organizations

Changing patriarchal organizations can be a complex and challenging process, but there are several strategies that can be effective in promoting gender equity and inclusiveness. Maintaining the patriarchal nature of research organizations leads can lead to negative research outcomes and missed opportunities for research to contribute to gender equality goals. It also leads to problematic narratives that instrumentalize women and to marginalize women, and gender research. However, we also find other more transformative discourses that, in troubling the drivers of gender inequality and promoting shared responsibility for change, reflect a deeper awareness of feminist scholarship and seek to address the power dynamics that are entrenched in such organizations. We use two case studies to explore initiatives shaping research and development organizations in the agriculture and food systems space to transform from patriarchal to transformative and equitable organizations.

6.3.1   Case Study 1: Building Accountability for Gender Equality in Food Systems Organizations: The Global Food 50/50[1]

The Global Food 5050 Team
The Global Food 50/50 accountability index assesses whether and how organizations are integrating gender and equality considerations in their work. It reviews the policies and practices of food systems organizations as they relate to two interlinked dimensions of inequality: inequality of opportunity in career pathways within organizations and inequality in who benefits from the global food system. The primary aim of the Global Food 50/50 Report is to encourage food system organizations to confront and address gender inequality both within their organizations and governance structures, and in their programmatic approaches across food systems. A second aim is to increase recognition of the role that gender plays in who runs and

---

[1] **Global Food 50/50 Study Team (listed alphabetically by last name):** *Global Health 50/50:* Kent Buse, Sarah Hawkes, Alex Parker, Sonja Tanaka; *International Food Policy Research Institute:* Jason Chow, Claire Davis, Lee Dixon, Hazel Malapit; *UN Women:* Carla Kay Kraft, Jemimah Njuki.

benefits from food systems that should work for everybody: women and men, including transgender people, and people with nonbinary gender identities.

As the world faces unprecedented levels of inequality in who benefits from the global food system, this report and accountability mechanism presents rigorous evidence on these inequalities and promotes accountability for change. Experience shows that public reporting of how organizations are doing on gender equality can push for changes, both internal and external. The Global Health 5050, a similar index in the Health Sector reported that between 2018 and 2022, these was a 25% increase in commitments to gender equality, a 27% increase in workplace policies on gender equality, and a 15% reduction in leadership bodies with fewer than one-third of women, as represented across 200 global health organizations. Organizations are assessed on nine variables across four dimensions (Table 6.1).

Commitment to Redistribute Power Public commitments to gender equality and to the redistribution of power are critical as a first step to addressing power dynamics in organizations. These commitments are reflected in the visions, missions, and core strategy documents of organizations. In 2022, only two organizations in the sample had not made a public commitment to gender equality. The commitments made however vary in the extent to which they can lead to fundamental change. For example 61% of organizations had made a commitment to gender equality for the benefit of all, while 35% had made commitments to gender equality, with a focus on empowering women and girls. Only 4% of organizations mentioned women and girls with no commitment to gender equality. There was change between 2021 and 2021, where 4% of organizations had no mention of gender equality or women and girls, in 2021, in comparison to 2022 where all the organizations either had made commitments to gender equality and or to women and girls. The definition of gender

Table 6.1 Global Health 5050; IFPRI and UN Women (2022)

Dimensions	Variables
1. **Commitment to redistribute power:** Organizational commitment to gender equality and an official definition of gender that is consistent with global norms.	Organizational public commitment to gender equality Organizational definition of gender
2. **Policies to tackle power and privilege imbalances at work:** Responsive policies that promote equality in attracting and retaining people, contribute to safe and respectful work environments, and are family friendly.	Workplace gender equality policy Workplace diversity and inclusion policy Board diversity policy
3. **Gender and geography of global food system leadership:** Outcomes in terms of gender balance in senior management, governing bodies and leadership, and the gender pay gap.	Gender parity in senior management and in the governing body Gender, nationality, education, and age of the executive head and chair of the governing body
4. **Addressing the gendered power dynamics of inequalities in outcomes** Global programs and monitoring that account for gender as a determinant of inequitable health outcomes.	Gender-responsiveness of global programs Sex-disaggregated monitoring and evaluation data

also matters. Definitions can exclude or include and can frame a problem and inform the solution. If we are to address the distribution of power across and within societies, institutions, and organizations, we need to understand gender as a social construct. Between 2021 and 2022, three more organizations had published their definitions of gender, with 59% of organizations defining gender in a way that is consistent with global norms (i.e., applying UN Women's definition). There are still gaps however as in 2022, more than a third of organizations did not define gender in their public strategies or policies.

Policies to Tackle Power and Privilege Imbalances at Work Evidence shows women remain underrepresented in the workplace and are particularly excluded from positions of power and decision-making, despite commitments to gender equality. In the 2022 report two-thirds of organizations had publicly available workplace policies with specific measures to advance gender equality. Specific measures included gender-responsive recruitment and hiring processes, mentoring, training, and leadership programs, targets for women's participation at senior levels, gender analysis and action in staff performance reviews and staff surveys, regular reviews of organizational efforts toward gender equality; and reporting back to all staff. Workplace inclusion and diversity policies are one of the common gender policies in the workplace. However only about half of the organizations in the report had workplace diversity and inclusion policies. One-quarter of the organizations were found to commit to diversity and inclusion but did not state how they were implementing that commitment. And 20% of the organizations did not have any commitment to diversity and inclusion.

Gender and Geography of Global Food System Leadership The distribution of gender in senior management reflects how an organization operationalizes its commitment to gender equality. It also provides insights into women's representation and voice in decision-making and leadership. While organizations are increasingly committed to gender equality and are putting policies in place, these good intentions are slow to be translated into the redistribution of opportunities and outcomes for women. The 2022 data shows that 35% of organizations had gender parity in their senior management teams and 15% had more women in senior management (56% or higher). Half of organizations had more men in senior management, including 30% that were composed of fewer than one-third women. Between 2021 and 2022 there was a slight reduction in organizations with fewer than one-third women in senior management. There were also improvements at the level of individual organizations. Seven organizations increased the number of women in their senior management in the past year, with two reaching gender parity.

Similarly, as compared to 2021, more women were represented on the governing bodies. This progress is obscured by 10 organizations that had fewer women in their senior management or governing bodies in 2022 than in 2021. The 2022 report showed that nationals of low-income countries continue to be grossly underrepresented holding just 3% of board seats and just 2% are occupied by women from low-income countries. By contrast, 71% of board seats are held by nationals of high-income countries. These countries represent only 16% of the global population. A high proportion of these seats were occupied by US nationals, who held 2 in 5 board seats (39%). Encouragingly, the proportion of women board chairs increased significantly from 24% in 2021 to 36% in 2022. Changing the status quo and redistributing power will not happen without intentional action. Affirmative measures to improve gender equality and diversity among board members, such as dedicated seats and board composition targets for underrepresented groups, are often necessary to institutionalize change. Board policies that contain specific measures are critical tools for realizing diverse and effective governance. Only a fraction of organizations had transparent policies to promote diversity on their boards. Policies with specific measures to promote diversity were found for 30% of organizations. This marks a 10% increase since 2021.

Taking a Gender-Responsive Approach to Improving Food Systems Gender norms play an important role in perpetuating inequities in global food systems across and within populations. Gender also influences how the food sector identifies, frames, and addresses these problems. In order to move toward gender-just and equitable food systems, organizations must adopt tactics that include gender-transformative planning, investment, and programming, as well as advocating for changes in the norms and power structures. In 2022, there was an increase in organizations with gender-transformative programmatic approaches from 60% to 70% and a decrease in the number of organizations with gender-blind approaches—with only one organization found to have genderblind programs. Sex-disaggregated data combined with gender analysis contribute to identifying disparities in food systems, including in access, consumption, and production. Sex-disaggregation of data is a means to hold organizations accountable for their commitments not only to equity but also to the delivery of effective interventions. Despite this only 3 in 5 organizations had publicly available policies committing to regularly sex-disaggregate their data.

6.3.2 Case Study 2: African Women in Agriculture Research and Development: Institutionalizing Gender Research in Agriculture Research for Development Institutions

Kenneth MachariaEdnah Kangogo, Michèle Mboo-Tchouawou and Susan Kaaria
AWARD was founded in 2008 as a career development program seeking to widen the pipeline of confident, capable, influential African women scientists in leadership in AR4D. AWARD initially targeted female scientists working in agricultural research in Africa, offering non-residential fellowships with a combination of activities that sought to establish mentoring relationships between early career scientists and senior, more accomplished scientists, and develop leadership capacity, plus provide science skills and gender training, among other knowledge building

interventions in ARD. The program sought to fix the leaky pipeline—a term used to describe the dropping off of women at various stages of their career growth within science, technology, engineering and mathematics fields. The Flagship AWARD Fellowship has benefited 714 fellows, reaching a total of 1773 individual scientists from 26 African countries when the fellows' mentors, and mentees are included. These scientists have helped catalyze gender responsiveness within their spheres of influence.

AWARD experienced remarkable success in empowering women scientists to grow in their visibility, confidence, influence, and professional capacity. But it quickly recognized that the pace of their progression was slow owing to the persistent systemic barriers and gaps in the awareness on gender-responsive processes in research and development initiatives, gender-related programming, and gender-sensitive work environments in their institutions. The Gender Responsive Agriculture for Research and Development (GRARD was established as a comprehensive solution to fill these gaps through serving as a channel for AWARD's support to African AR4D institutions in their quest to be more gender responsive in their internal and external processes. Funding for GRARD came from the Bill & Melinda Gates Foundation and the United States Agency for International Development.

GRARD focuses on supporting African AR4D institutions to address two main challenges:

- The lack of external gender responsiveness, which is the failure to recognize and prioritize the distinct needs of different groups, and especially women, in research agendas and processes, and
- The lack of internal gender responsiveness, which is the absence of gender diversity, particularly the underrepresentation of women scientists and practitioners, in agricultural leadership and roles.

External and internal gender responsiveness are both crucial for achieving gender equality and effectiveness in AR4D institutions. External gender responsiveness in research encompasses the distinct needs and priorities of a diversity of both men and women across the entire agricultural value chain. Gender-responsive research can significantly enhance agri-food systems for Africa's economic growth by designing research agendas that promote sustainable and inclusive rural transformation. By focusing on innovation that addresses the constraints faced by marginalized African farmers, particularly women, gender responsiveness can maximize the impact and efficiency of AR4D institutions in Africa. Internal gender responsiveness includes aspects related to internally orientated strategies and interventions to support gender integration in staffing, human resource policies, the vision and mission, and the organizational culture. The pertinent human resource policies include those related to diversity; equality and fairness in hiring, compensation, and advancement; mentoring; promotion; and professional development. Other essential elements are well-resourced gender strategies and action plans and gender-sensitive systems for data collection and for assessment of key performance metrics related to gender.

The selection of the institutions for piloting GRARD involved extensive consultations with key stakeholders and partners in the AR4D sector. Eight institutions were chosen, including national agricultural research centers and universities, based on their longstanding relationship with AWARD and their commitment to prioritizing gender responsiveness. Two of the institutions were treated as the leading recipients of the support owing to the commitment shown by their leaders to gender advancement. The other institutions were to benefit from regional level activities initially, with the expectation of their deeper engagement in subsequent phases.

AWARD's Conceptual Framework for GRARD AWARD and Gender at Work (forthcoming) have developed a conceptual framework for nurturing transformative gender-related change in African AR4D institutions. Figure 6.1 provides a comprehensive overview of the different factors and pathways that can contribute to such transformation. Institutions can use this framework to map their current strategies and gaps and identify the priorities for future activities in support of gender equality in their context.

AWARD's framework directs attention to the internal and external factors that influence how gender, diversity and inclusion are addressed in research institutions. The outer circle of the framework in Fig. 6.1 contains the five domains on the internally orientated strategies and interventions to support gender inclusion in staffing and human resources policies, the vision and mission of the institution, and data disaggregation, and the supporting requirements in the form of a well-resourced gender strategy or action plan and an inclusive organizational culture. The inner circle focuses on the domains of the institutions' external influence on how gender is addressed in research. The four areas for action are the institution's leadership support for gender research, recognition of gender analysis as a key competency for all researchers, inclusion of gender in research processes, and incorporation of gender in research dissemination.

The process to operationalize the conceptual framework among the GRARD institutions involved the following steps: (i) Conducting a participatory gender audit or needs assessment: This step analyzed the integration of gender in both the institution's internal and external dimensions including the internal factors, that is the institution's policies, structures, and practices and how they enabled or constrained gender integration into the design and implementation of research. Also assessed were elements such as staffing policies for human resources management and career and capacity development, the work environment and if family friendly policies and practices existed, and organizational culture. For the external factors, such as how gender was addressed in the institution's projects and programs, and how that could be strengthened, plus the institution's capacity for gender-responsive research and how it could be enhanced, including the interlinkages between the external and internal processes, focusing on the key program functions that could affect the integration of gender in the institution's research programs and the functions that would need to be addressed to improve the results of the institution's development work, (ii) Designing the institutional gender action plan: The gender audit led to the development of an institutional gender action plan whose purpose was to provide the strategic and practical direction for the institution's internal and external gender integration. The action plan also clarified the changes expected to occur and the

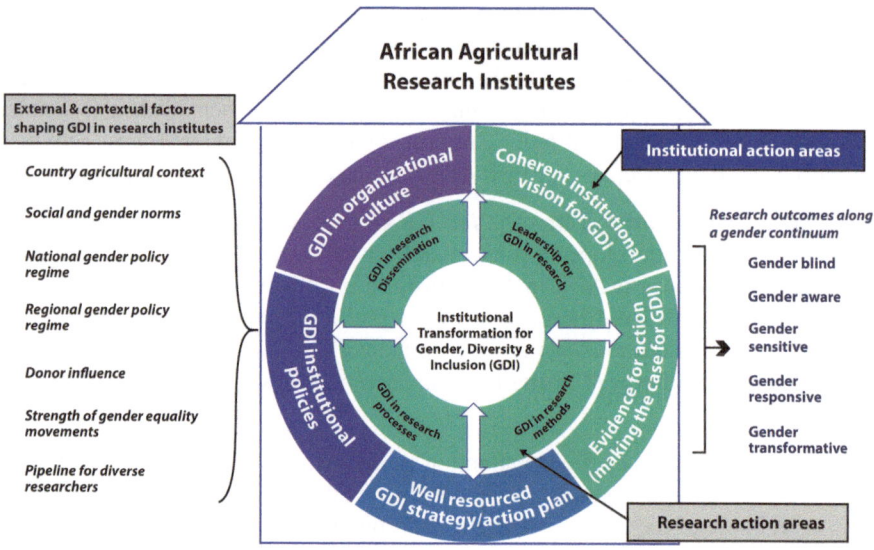

Fig. 6.1 Conceptual framework for transformative change. (Source: AWARD & Gender at Work (forthcoming))

mechanisms through which they were expected to happen. The gender action plan identified the entry points for the different types of strategies and prioritized their implementation at the enabling environment, institutional and individual levels (iii) Implementing the institutional gender action plan through capacity development: Interventions in the gender action plan needed to be tailored to the circumstances and priorities of the individual institutions. The participatory nature of the process allowed the prioritization of the identified internal or external GRARD strategies or both. The activities included capacity development for institutional leaders, research directors, human resources managers, gender focal points, and other staff as needed, to raise awareness about the value of and the methods for advancing gender responsiveness. Training workshops helped participants to learn how to lead the change process in their specific institutional gender action plans.

Example of the Federal University of Agriculture (FUNAAB), Abeokuta, Nigeria FUNAAB in Ogun State in southwest Nigeria sought from its inception in 1988 to serve national development through agricultural and rural development research. It is a specialized institution with a mission to conduct agricultural research and rural development studies for sustainable national development. FUNAAB's mandate is tripartite in nature, comprising of teaching, research, and strong community engagement comprising extension services with adopted villages and other research institutions. FUNAAB consists of 10 colleges, 45 departments and 30 centers/directorates. Through its research engagement with national and international stakeholders, it is known throughout Africa and in other continents as a champion in food security, environmental resource management and agricultural research. FUNAAB first engaged with AWARD in 2008 when five of its staff members participated in the second cohort of the Flagship AWARD Fellowship, with two of them as fellows, two as the fellows' mentees, and one as a mentor. Since then, 35 FUNAAB staff members have been involved in AWARD fellowship programs and have benefited through the career acceleration program to rise to various positions of leadership and influence in their institution, including as the Deputy Vice-chancellor for one of the mentors. This group, which forms a critical mass of AWARD alumni, benefited from various types of training on the importance of gender in their research work, which saw them champion the move towards gender inclusiveness in their institution. FUNAAB commenced the creation of a gender strategy in 2016. FUNAAB was among the institutions selected to be part of the GRARD pilot program and was involved in various conceptualization workshops, which led to its signing of a memorandum of understanding (MOU) with AWARD in 2018.

Gender Audit and Institutional Needs Assessment With the support of AWARD, FUNAAB conducted its own institutional needs assessment, which was a study involving 115 teaching and non-teaching staff, of whom 55 were women. The research employed questionnaires and focus group discussions consisting of 21 staff, among whom 11 were women. The aims were to (1) identify the gains and gaps in the institutional mechanisms that needed addressing for gender responsiveness in AR4D, (2) assess the capacity needs and define the innovative mechanisms to spearhead the implementation of gender responsive AR4D in the institution, and (3) review and upgrade the gender policy to meet the needs of all FUNAAB stakeholders. The needs assessment study found FUNAAB to have significant gender responsiveness achievements but also some gaps in institutional mechanisms, highlighting the type of support needed by the institution. In terms of achievements, the university had set up a gender unit along with a gender committee to guide its activities, had a draft gender policy, had introduced gender content in various courses within the university curriculum; and had made deliberate moves towards opening up leadership roles for women, including at the levels of deputy vice chancellor, directors, deans, and heads of departments. Several gaps were however identified including a lack of awareness by some staff members of the existence of the draft gender policy, the gender unit or the staff involved in them, inequity in benefits from the training programs, and a limited number of women taking up leadership positions within the university.

Designing the Institutional Gender Policy The assessment of the institutional processes and its findings and recommendations paved the way for a review of the gender policy document. In 2021, FUNAAB's senate approved the review of the *Gender mainstreaming guidelines*, and a six person Gender Policy Review Committee was constituted to review the 2016 *Gender mainstreaming policy* for updating. The policy, which had stalled since 2016, was completed and ratified by the university senate and launched by the Vice Chancellor during a stakeholder meeting organized by AWARD. AWARD's support for the review process was through a gender expert who was also a key actor in the institution's needs assessment. The *Gender mainstreaming policy* was updated and the new version was launched in December 2022. It was a product of reflection, consultation, and data analysis over several months to determine the specific aspects for consideration and the critical entry points for promoting gender responsiveness in the policies and practices at FUNAAB. FUNAAB's gender policy recognizes the need for structural changes to strategically elevate the institution's position in the gender and agricultural policy agenda. The new perspectives in the revised gender policy provide a platform for FUNAAB to engage further with its key stakeholders on the relevance and prioritization of gender responsiveness in the institution's policies and practices.

Institutional Capacity Development to Support Implementation of the Gender Policy

AWARD conducted a series of capacity development interventions at FUNAAB, targeting both academic and non-academic staff from junior levels to senior managers. The capacity development interventions resulted in the training of over 150 staff members, who were mostly senior managers, researchers, and administrators. The interventions included the institutional mentoring orientation workshop, a progress monitoring meeting, the scientific writing and publishing skills course, a half day training on gender for FUNAAB senior management, and the leadership and assertiveness course:

- **Mentoring orientation workshop:** The entry point for institutionalizing mentoring was via a program that adopted the AWARD mentoring model. The first mentoring orientation workshop, which was conducted by AWARD, attracted 24 mentoring pairs. FUNAAB took over the running of the program after that.
- **Leadership and science skills training:** Training activities were conducted for building leadership capacity and science skills in relation to gender. The leadership training specifically targeted emerging leaders in the institution as a way of preparing the next generation of influential scientists for the institution's growth. The science writing training focused on building science and proposal writing skills so that junior and mid-level scientists could gain visibility through publishing and by leading research projects for which they had sought funding. In addition, AWARD organized a follow-up science clinic to evaluate the progress made and provide additional support to all staff members who had benefited from the science training.
- **Gender in agriculture training:** AWARD and FUNAAB's recognition of the need to build awareness on gender among the senior management and other staff members led to AWARD conducting various gender awareness training sessions during the period of the intervention that had over 100 participants, including staff members from partner institutions.

6.4   Key Elements for Transforming Patriarchal Organizations

While the above examples provide some elements set of actions to transform organizations, changing patriarchal organizations requires a comprehensive and sustained effort. This includes organizations prioritizing and promoting diversity and inclusion to create an environment that values and respects differences in gender, race, ethnicity, age, and other characteristics. This can involve implementing policies and practices that support gender equity, such as offering parental leave and flexible work arrangements, addressing gender bias in hiring and promotion, and providing training on diversity and inclusion. Another key element is encouraging women's leadership and ensuring there are policies in a place that remove bias. This can involve workplace policies, mentoring and sponsorship programs, leadership training, and creating a culture that values and supports women in leadership positions.

Addressing gender bias is critical, including addressing discrimination in all its forms, unconscious bias in hiring and promotion, gender-based harassment and discrimination, and gender pay gaps. This can involve conducting diversity audits, creating policies and practices that promote gender equity, and providing training on unconscious bias and gender sensitivity. Removing hierarchies in organizations is critical to ensure all forms of research and all processes are accorded equal value. Placing equal importance for example between the social sciences and the biophysical sciences, between the science technology mathematics, and the arts and humanities as supplemental forms of science and not as one in service of the other ensure that priority is given to all.

A last key element from these case studies is monitoring and accountability, through transparent commitments, reporting on gender equity and inclusion through clearly set transformative goals and benchmarks for gender equity that go beyond counting numbers to look deeper into organizations, their leadership, their policies, and the transformative nature of their outcomes. For the Global Food 5050, public presentation of how organizations fare on different indicators in a form of "name and shame" is expected to lead to better performance by organizations across the different elements of gender equality.

While these elements may be useful, they most often fall short of the dismantling the structural underlying building blocks of patriarchal organizations. Talbot (2002) in analyzing the patriarchal and gendered dynamics of sports organizations observes that it is assumed if women can learn new behaviors such as leadership skills, how to be more assertive, or confident, that they will find their way in the organization. The way the organization itself is structured, managed or organized is not seen as the barrier. What the case study from AWARD shows is that in addition to having capable and confident women leaders, the policies and structures of the organization must also change.

Research and development organizations which claim to embrace the values of equity and inclusiveness cannot afford to ignore structural inequalities or refuse to acknowledge their own part in perpetuating them. Feminist scholars such as Halford (1991) have argued about the ability to bring structural, and indeed feminist social change into organizations of state and especially given the ways in which state actions often reflect and reinforce the dominance of men. Buswell and Jenkins (1994) argue that even the best-intentioned equal opportunities policies fail to address not only structural inequalities but also the role that organizations themselves play in maintaining gendered hierarchies.

Dismantling patriarchy internally and in the wider world is only possible if first, there is a shared understanding of its nature and how it operates with other structural oppressions to organize society. It requires challenging gender norms and stereotypes that perpetuate gender-based power imbalances and promoting more diverse and inclusive leadership and structures within organizations.

References

Acker J (2006) Inequality regimes: gender, class, and race in organizations. Gend Soc 20(4):441–464

Beechey V (1979) On patriarchy. Fem Rev 3(1):66–82

Beintema N (2017) An assessment of the gender gap in African agricultural research capacities. J Gender Agric Food Secur 2(1):1–13. https://doi.org/10.19268/JGAFS.212017.1

Benschop Y, Verloo M (2006) Sisyphus' sisters: Can gender mainstreaming escape the genderedness of organizations? J Gender stud 15(1):19–33

Buswell C, Jenkins S (1994) Equal opportunities policies, employment and patriarchy. Gender, Work & Organization, 1(2):83–93

Espeland W, Stevens M (2007) A sociology of quantification. European Journal of Sociology/Archives européennes de sociologie 49(3):401–436

FAO (2023) The status of women in agrifood systems. Rome https://doi.org/10.4060/cc5343en

Farh JL, Cheng BS (2000) A cultural analysis of paternalistic leadership in chinese organizations. https://doi.org/10.1057/9780230511590

Fogarty MP, Allen I, Walters P (1981) Women in Top Jobs, London: Heinemann Education Books

Global Health 50/50, the International Food Policy Research Institute, UN Women (2022) Hungry for gender equality: the global food 50/50 report 2022, Washington, D.C. https://doi.org/10.56649/WIQE2012

Guba EC (1990) The alternative paradigm dialog. In: Guba EC (ed) The paradigm dialog. Sage, Newbury Park, pp 17–27

Halford S (1991) Feminist change in a patriarchal organisation: the experience of women's initiatives in local government and implications for feminist perspectives on state institutions. Sociol Rev 39(1_suppl):155–185. https://doi.org/10.1111/j.1467-954X.1991.tb03359.x

Hart J (2016) Dissecting a gendered organization: implications for career trajectories for mid-career faculty women in STEM. J High Educ 87(5):605–634. https://doi.org/10.1080/00221546.2016.11777416

Hartman M (1979) A descriptive study of the language of men and women born in Maine around 1900. In Betty-Lou Dubois & Isobel M. Crouch (eds.), 81–90

Hedges LV (1987) How hard is hard science, how soft is soft science? The empirical cumulativeness of research. Am Psychol 42:443–455. https://doi.org/10.1037/0003-066X.42.5.443

Jackson JC, Payumo JG, Jamison AJ, Conteh ML, Chirawu P (2022) Perspectives on gender in science, technology, and innovation: a review of sub-Saharan Africa's science granting councils and achieving the sustainable development goals. Front Res Metr Anal 7:814600. https://doi.org/10.3389/frma.2022.814600. PMID: 35480783; PMCID: PMC9035601

Kaaria S et al (2016) Rural women's participation in producer organizations: An analysis of the barriers that women face and strategies to foster equitable and effective participation. J Gender Agric Food Secur 1(2):148–167

MacMynowski DP (2007) Pausing at the brink of interdisciplinarity: power and knowledge at the meeting of social and biophysical science. Ecol Soc 12:20–33

Mbo'o-Tchouawou M, Musembi L, Beyene TA, Kamau-Rutenberg W (2019). Gender and leadership in Africa: exploring the nexus, trends, and opportunities. https://doi.org/10.2499/9780896293649_03

Njuki J, Bukachi S (2021) Gender considerations and practices for transforming tertiary agricultural education in Africa. In: Transforming tertiary agricultural education in Africa. CABI, Wallingford UK, pp 182–197

Ramsay K, Parker M (1991) Gender, bureaucracy and organizational culture. Sociol Rev 39(1_suppl):253–276. https://doi.org/10.1111/j.1467-954X.1991.tb03363.x

Scott I (1998) The Bureaucratic Transition. In Institutional Change and the Political Transition in Hong Kong (pp. 158-179). London: Palgrave Macmillan UK

Talbot M (2002) Playing with patriarchy: the gendered dynamics of sports organizations. In: Scraton S, Flintoff A (eds) Gender and sport: a reader. Routledge, London, pp 277–292

Timmermans S, Epstein A (2010) A world of standards but not a standard world: toward a sociology of standards and standardization. Ann Rev Sociol 36:69–89

UNESCO, Schlegel F (eds) (2015) UNESCO science report: towards 2030. UNESCO Publ, Paris, p 794. (UNESCO science report)

van Eerdewijk A (2014) Gender mainstreaming: Views of a post-Beijing feminist. In The Palgrave Handbook of Gender and Development: Critical Engagements in Feminist Theory and Practice (pp. 117-131). London: Palgrave Macmillan UK

Walby S (1998) Heorising patriarchy. Sociology 23(2):213–234

Weber M (2016) Economy and society. In: Democracy: a reader. Columbia University Press, New York pp 247–251

Witz A (1990) Patriarchy and professions: the gendered politics of occupational closure. Sociology 24(4):675–690

Witz A, Savage M (1991) The gender of organizations. Sociol Rev 39(1_suppl):3–62. https://doi.org/10.1111/j.1467-954X.1991.tb03355.x

Zippia (2023) Women in leadership statistics. Facts on the Gender Gap in Corporate and Political Leadership. 8 June 2023. https://www.zippia.com/advice/women-in-leadership-statistics/

Chapter 7
From Power to Women's Empowerment: The Missing Links

Vivian Polar ⓘ **and Nigel Poole** ⓘ

Abstract This chapter explores the instrumentalization of women's empowerment in agricultural research for development, with particular attention on critically examining how the concept of empowerment has become understood as an externalized process that can be bestowed on women through production-oriented interventions. The chapter explores multiple manifestations of power and depicts their occurrence through experiences of women and men farmers in the Andean region. It analyzes how the use of empowerment has deviated from building agency and disrupting power dynamics, highlighting the need for a feminist and transformative conceptualization and operationalization of empowerment in the agricultural sector.

7.1 Introduction

When researchers, policy makers and development practitioners in the agricultural sector use the word empowerment to refer to research and development goals, the underlying assumptions may vary significantly from the way the concept of empowerment evolved through philosophical, social, political, and feminist thinking.

Analyzing the adequacy of multiple accounts of power is no easy task. If we focus on feminist conceptions of power, we see that many of them have been reconstructed out of debates on critical topics such as pornography, motherhood, marriage, sexual harassment, care, and equality (Allen 1998). However, few accounts have analyzed the conceptualization and manifestation of power and empowerment in technically entrenched masculine topics such as agricultural research and development.

V. Polar (✉)
International Potato Center, Lima, Peru
e-mail: v.polar@cgiar.org

N. Poole
SOAS University of London, London, UK
e-mail: np10@soas.ac.uk

© The Author(s) 2025
J. Njuki et al. (eds.), *Gender, Power and Politics in Agriculture*,
https://doi.org/10.1007/978-3-031-60986-2_7

This chapter will walk us through the evolution of the term empowerment, often questioned for its instrumental use. It will start by analysing the concept of power as it emerges from philosophical and political thinking, and how it relates to concepts introduced by feminist scholars. Finally, using diverse lines of thinking, a holistic definition of empowerment is presented to visualize how the agricultural sector has focused its attention on specific aspects that address an incomplete image of empowerment, divorced from its political and transformative nature.

7.2   Disentangling the Definition of Power

Modern notions of power began in the sixteenth and seventeenth centuries. Nicolo Machiavelli in his famous book "The Prince" describes power as a resource and analyses the strategies and management of power (Machiavelli et al. 2006). A century later Thomas Hobbes, in "The Leviathan" represents the causal thinking of power as a hegemony and conceptualizes it as the means to obtain some future apparent good; classifying it as inherent and acquired (Hobbes and Tuck 1996). These two contrasting representations, the first focuses on the mechanisms of power and the second visualizes it through a moral perspective, continue to be the two main routes of thought about power (Clegg 1989).

In the latter half of the Twentieth Century, the definitions of power advanced focusing on it as a relational phenomenon, reflecting the relationship between the powerful and the powerless. Max Weber linked power with concepts of authority and law, visualizing power as a factor of domination which he defines as *'the probability that one actor within a social relationship will be in a position to carry out his own will despite resistance...'* (Weber et al. 1979). This definition was the cornerstone of an interpretation of power as 'power-to', meaning 'power to accomplish some purpose'. Robert Dahl located the discussion of power inside the boundaries of a community. Within this framework, power is exercised by particular individuals to prevent others from doing what they would rather do, or to follow the preferences of those who possess the power (Dahl 1971). This perception of power is the origin of what later has been called 'power-over'. These two definitions of power (power-to vs power-over) have been central to an on-going debate amongst social scientists.

A series of models and theories have emerged to explain the nature and occurrence of power. Three dimensions are often presented to explain the different ways in which power is manifest. The first manifestation of power, also referred to as the "overt face of power", is an intuitive idea (Dahl 1971) that considers action over decision making. The second dimension or the "covert face of power", touches on the prevention of decision making (Bachrach and Baratz 1962). The third dimension is what Lukes calls the 'latent dimension' that refers to the implantation in people's minds of interests contrary to their own good (Lukes 2005). In the same line of thinking, John Gaventa in his power cube recognizes three degrees of visibility of power, distinguishing between visible, hidden and invisible manifestations of power (Gaventa 2006). This three dimensional perspective is challenged by Michel

Foucault who systematically rejects the existence of power as a source from which actions stem and pictures only an infinite series of practices, thus decentralizing the concept and extrapolating it from sociology to all fields of the social sciences and humanities (Foucault and Faubion 2002). Even though the roots of power as a concept are grounded in political theory and philosophy, its importance has gradually been established in contemporary sociological discourse.

More recent definitions of power and their related theories have continued to evolve in search of models that explain the way power processes are effected in society. For example, Gaventa's model of power and powerlessness that emerges from Lukes's tri-dimensional view and seeks to explain situations of social inequality, uncovering the direct and indirect ways in which social powerlessness is created and maintained (Gaventa 1980). Giddens' theory of structuration is a dialectic vision of power where all human actions are at least partly predetermined by the varying rules of a specific context (Giddens 1984). Both lines of thinking show an evolution of the concept of power. The debate reflects new dimensions of an analysis that began in the political sciences, but has since entered vigorously into other social sciences.

7.2.1 Mainstream Definitions of Power

There are two main models or definitions of power with a clear division established between 'power-to' and 'power-over'.

7.2.1.1 Power to

'Power-to' is defined as the capacity to have an effect. It is about agency and is regarded as generative or productive power which creates new possibilities and actions (Rowlands 1997). The definition of 'power-to' views power as ever-expanding energy (Hartsock 1985; Parsons 1963). It uses an image of human development and considers power to be infinite and innocuous in its effect over others. The danger with this perspective is that it can suggest that power is a personal attribute (Nelson and Wright 2001), thus placing responsibility for powerfulness and powerlessness on the individual. This definition of power informs the capability approach of Amartya Sen, who asserts that people are not free when they do not have power to make choices about their lives (Sen 1995). Therefore, 'power-to' focuses mainly on behaviour (Lukes 2005) of decision making or its prevention (Bachrach and Baratz 1962). Building on the same argumentation, concepts of 'power-with' and 'power-from-within' emerged to describe the power phenomenon from collective and internal perspectives; building on the experience of women as mothers and caregivers, giving way to power definitions that reproduce transformative growth for oneself and for others (Held 1993):

'Power-with' is a collective ability based on relationships of reciprocity between members of a group (Follet 2003). It is regarded as collective action in response to powerlessness (Eyben 2005), reflecting a sense of the whole being greater than the sum of the individuals. Hence it is a positive-sum phenomenon (Rowlands 1997). This definition of power highlights solidarity and a collective ability based on receptivity and reciprocity within the group (Allen 1998). Power-with results from individuals organizing and acting together on common concerns (Gammage et al. 2016).

'Power-from within' reflects the inner strength of every individual, and is based on self-acceptance, self-respect (Rowlands 1997) and self-worth (Eyben 2005). In feminist thinking this concept is visualized as positive, life-affirming, and an empowering force antagonistic to power understood as domination.

'Power-through' captures an involuntary manifestation of power that operates at the intersect of power to and power over. It is a distinctive and personal manifestation of power but mediated through the existence of others as individuals, communities and values (Galiè and Farnworth 2019). Power through takes a step towards making sense of complex and multifarious power relations by evidencing how elements of domination and culture manifest to shape individual actions and experiences.

7.2.1.2   Power-Over

'Power-over' is based on a different image. While 'power-to' reflects on an infinitely expanding and innocuous process, 'power-over' pictures a closed system of power fluctuation, a zero-sum phenomenon where one gains power at the expense of another, and where power relations are coercive. It is perceived as controlling power to which the response may be compliance, resistance or manipulation (Rowlands 1997). This definition includes a behavioural component, yet it is also a critique of the behavioural focus since its main characteristic is the analysis of observable and latent conflict. It illuminates the systematic ways in which power is perpetuated and exercised to prevent conflict (Gaventa 1980; Lukes 2005).

7.2.1.3   When Power-to Meets Power-Over

Definitions of power highlight distinctive features and manifestations that aid their classification. However, the way they operate and interact in real life is complex and intertwined, as illustrated in the following cases. Case 7.1 describes an intervention that sought to empower men and women farmers by building individual and collective agency in a community in southern Bolivia. In Case 7.1 we identify how the intervention sought to build **power-from-within** through enhancing individual capabilities in project management and self-esteem. The intervention developed **power-with** by strengthening the local farmer organization, and fostered **power-to** by building capacity for collective negotiation to access technical assistance. Case 7.1 is a good example of how all three types of power operate. Intense negotiations by the farmer organization show clear evidence of the exercise of power-to, the strength of the farmer organization is a reflection of power-with because as a team their voices are heard and motivate government officials who would otherwise not respond to individual requests. Farmers exercise power-from-within when they individually decide to stand and reject the equipment that does not meet their needs.

While Case 7.1 shows clear evidence of the expansion of different forms of power, there is a barrier that limits the achievement of positive outcomes. **Power-over** in this case is evident in the actions of the local government that holds total decision-making power about how funds are ultimately allocated. Their attempt to deliver sub-optimal equipment in a public space was a form of coercion where they tried to swing public opinion in their favour and pressure farmers to accept the equipment.

Case 7.1 "We Will Not Take This Equipment": Defining the Terms of Service Provision (Yacuiba—Bolivia)

In an agricultural development project in the Chaco region of Bolivia, the service provider hired by the local government had committed to providing technology advisory services and equipment for maize processing to farmer organizations. Men and women farmers participating in the project engaged in discussions about technical characteristics of the equipment and the nature of advisory services. In parallel the farmer organization received capacity building in Participatory Monitoring and Evaluation (PM&E) as a mechanism to foster individual and collective empowerment. As the agricultural project unfolded, farmers applied a simple (PM&E) approach and used the information to alert the service provider and the local government about strong dissatisfaction from farmers.

Technical staff reported that although capacity building events had good participation and farmers were satisfied with the information provided, the negative overall assessment was because the equipment committed was not being delivered. The service provider explained to farmers that the local government had not yet provided the funding for the equipment. In response, a delegation from the farmer organization decided to escalate the complaint and engaged in several meetings with local government officials to request payments for the service provider and the delivery of the equipment.

The local government had unilaterally decided to re-allocate the budget for equipment to other activities but under the intense pressure from farmer groups, government officials negotiated with the service provider and other donors, purchased equipment for maize processing and organized a big event including the media to deliver the equipment. However, during the event, farmers rejected the equipment publicly because they realized it did not meet the technical standards required and previously agreed.

Ultimately, despite the intense negotiation and pressure from the farmer organization, they were never able to access equipment that met the standards required. The PM&E skills, intended to empower farmers, built project management skills, individual self-esteem and collective capacity to negotiate but the lack of positive outcomes left them with a lower sense of empowerment after the intervention (Polar 2013).

Based on participant observation and results from the implementation of Participatory Monitoring and Evaluation in an applied agricultural innovation project in Yacuiba—Bolivia (Fernandez et al. 2012).

In addition to the observable power dynamics in Case 7.1, there are also multiple other layers of power-over being exercised. The farmer organization was essentially led by medium and large-scale male farmers. The processing equipment demanded was more suitable for medium scale farmers and not as efficient for smaller scale farmers. Very few women were present in negotiations and their presence was used to strengthen the collective's image. However, women were never asked if the equipment delivered would fit their needs. Most women were actually small-scale producers more interested in the capacity development and in simple and manual equipment, rather than the equipment demanded by men, which required an external power source. Their preferences were not considered in the definition of technical specification, but their presence was used during delivery negotiations. This shows how despite being part of "power-to" and "power-with" dynamics, the underlying structures of power-over that shape the agenda of collective action may dilute the possibilities of positive outcomes for women.

7.2.2 Power in Postcolonial, Decolonial and Analytic Feminism

If we understand that our conceptions of power are themselves shaped by power relations (Lukes 2005) and that differentials of power come already embedded in culture (Yanagisako and Delaney 1995), we must delve into the realm of postcolonial and decolonial theory that questions if the oppressed can actually speak (Spivak 1988), or exercise any type of power while caught between imperial discourse and patriarchal tradition. Furthermore, postcolonial feminism also questions the overly simplistic understanding of power and oppression as reductive, homogenizing class, race, religion, and daily material practices of women in the Third World to create a false sense of the commonality of oppressions, interests, and struggles between and among women globally. Post-colonial feminism builds on the work of Quijano and Lugones to analyze the coloniality of power as a system strictly characterized by sexual dimorphism (Lugones 2007, 2010; Quijano 2019) where gender becomes another element of oppression, and a mechanism to exercise the agenda of patriarchy, capitalism and the state (Apffel-Marglin and Sanchez 2002).

In a similar path, Cudd uses the framework of rational choice theory to analyze oppression and power, conceptualizing oppression as normative or structural. By appealing to a structural theory of choice, Cudd disentangles oppression from assumptions about the individual's capabilities. Agents behave rationally, choosing actions that maximize their utility but, in a context where individual choice is constrained within socially structured payoffs (Cudd 2006).

Case 7.2 shows clearly how Spivak's questioning the effective role of someone who has been marginalized or oppressed -a subaltern -operates in a patriarchal and postcolonial context. Despite the presence of technical service providers dedicated to the dissemination of agricultural technologies, women do not ask questions about the information being shared, because the forum is one of men speaking to men. Their social position as women and the dominant language, both inherited from colonial rule limit women's access to technological alternatives. Their capabilities to exercise choice of agricultural innovations are constrained.

Case 7.2 Invisible Twice: A Woman and an Aymara Speaker
In Jacopampa— on the high Altiplano of Western Bolivia, women and men potato farmers received technical assistance and adopted technologies for seed production. During focus group discussions women were consulted about the information received, the technologies they used most and the technologies they decided not to use. Follow up interviews further explored the reasons for not using some specific technologies such as the bio-insecticide (Matapol-Plus) referred to below. An Aymara woman, approximately 40 years old, mentioned:

> I have Matapol-Plus but I don't use it. I received Matapol-Plus as a prize at the local fair two seasons ago but I did not use it because I don't know if it's toxic or not and also I don't know how to use it and what it is for. The technician came to the community to talk about it and showed pictures of potato storage but it was presented in Spanish and I did not understand.

During capacity building events women would sit at the back. Some of them attended their small children, and rarely asked questions or volunteered to participate in practical exercises led by technical staff. Many of them spoke limited Spanish and could not read. In this context agricultural technology options were twice invisible, once due to gender roles and socially accepted norms that limited women's engagement in capacity building, and the second time due to the presentation of technology in verbal and written form in a language in which the women had limited understanding.

Based on a qualitative study conducted in the context of the IssAndes project in Bolivia, Ecuador and Peru (Polar et al. 2015).

7.2.3 "Power Feminism"—Post-feminism

"Power feminism" is a school of thought that seeks to recapture progressive politics, reorienting discussions within feminism away from the excessive attention to women's victimization, to one that highlights women's newfound power (Caputi 2013; Hains 2009; Wolf 1994). Power feminism may also be considered to fall under the umbrella of post-feminism, that draws on the first and second waves of feminism but rejects their most provocative challenges such as those linked to critiques of capitalism and class privilege, as well as concepts of patriarchy and collective action (Hains 2009; Vavrus 2002).

The dichotomy of power feminism vs victim feminism described by Naomi Wolf in the power feminism approach also emphasizes individualism, conceiving it as a binary opposed to collective action (Hains 2009). The emphasis on individualism and triumph creates a shadow that enhances the gap between those who can experience enhanced individual capabilities and those who operate in collective environments and who continue to experience oppression and oppressive structures such as those described in Case 7.2.

Poverty itself creates an environment that shapes priorities differently. For example, in a dialogue of women's movements in Bolivia, Alexia Escobar, a Bolivian anthropologist and communicator asserts that: "Feminist groups vindicate as a main topic the right to decide over our own bodies while sisters from indigenous and farmer organizations raise malnutrition as their main issue. We've had difficulties to advance in agreements because we essentially have different codes"(Wanderley 2010).

7.2.4   Power and the Spheres of Life

There are varied spaces, places and domains where power is exercised in its multiple forms. However, to facilitate analysis we will connect the different definitions of power to three main spheres of life or domains of action: individual, collective and structural. Figure 7.1 shows a graphical understanding of power definitions and the spheres of life.

Power-over, often perceived as domination, is strongly linked to the structural sphere that relates to legal frameworks, institutional processes and mechanisms, culturally accepted structures, norms and other formally and informally established structures of power (Polar 2013). However, power-over does not exclusively

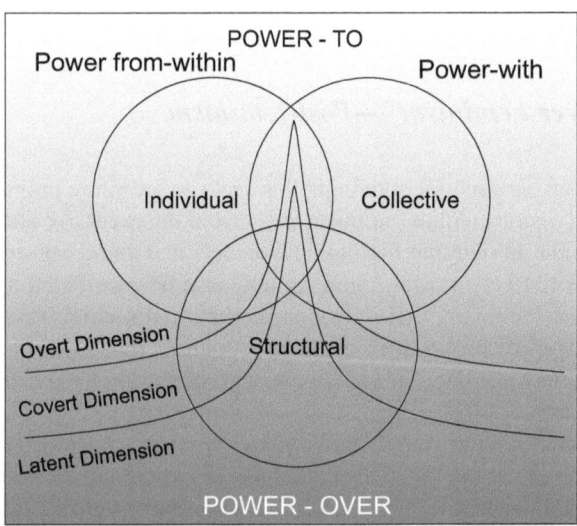

Fig. 7.1 Definitions of power and their interaction with different spheres of life. (Source: From Polar, 2013)

manifest in the structural sphere, it expands to influence directly the individual and collective spheres (notice the intersecting areas of the spheres) and indirectly through the action of the structural sphere (notice the shade of power-over reaching out to all spheres in different degrees). The different levels or dimensions of power-over also change from the latent dimension to the covert and overt dimensions connecting with all spheres of life in the process. It is in this complex of interactions where insights from post-colonial, de-colonial and analytical feminism need to be included.

The concept of power-to, on the other hand, is closely related to the individual and collective spheres. Power-with is linked to the collective sphere, reflecting collective action and solidarity. In parallel, the individual sphere is better explained by the power-from-within concept, as shown in Case 7.3.

Case 7.3 exemplifies how gender norms and roles that operate in the context of native potato cultivation and conservation shape individual and collective experiences, ultimately shaping emerging social structures. At the individual level women's reproductive roles at home, reinforced by local beliefs and cultural traditions, restrict their mobility and limit their involvement in some agricultural practices while strengthening their productive role in seed management and diversity conservation. However, it is this same reproductive role that has shaped culturally accepted norms about mobility and participation in public spaces, that ultimately influences women's collective experience and shapes emerging structures such as the AGUAPAN Association.

The underlying question is: how can institutional and organizational structures be designed to bridge or resist pressure from persistent gender norms and roles that limit women's participation? While incorporating quotas to enhance women's participation is a good starting point, Case 7.3 shows that it is not only about having a space to participate or the recognition, but also about having the time and mobility to engage. In the context of generalized poverty where women are responsible for most reproductive labor, such as those experienced in the high Andes where native potatoes are produced, a push towards further productive demands on their time and labor may enhance disparity and affect their overall sense of wellbeing. In such cases an empowerment and development agenda must centrally address the transformation of social relations of production.

7.2.5 Power and the Driving Forces of Change: Agency and Structure

The relationship between individuals and society or agency and structure is one of the central and contested issues in social sciences. The concepts of agency and structure are organized around two axioms: (a) individuals (human beings and organizations) act purposefully to transform the society in which they live; and (b) social

Case 7.3 A Mother to Her Children, a Mother to the Seed

A study on gender roles in native potato diversity management in highland communities of Peru showed that a traditional and quasi-religious view of women's reproductive roles flows across the different spheres of life and shape the way native potato diversity is managed and conserved.

In the Peruvian highlands, tradition asserts that women are not supposed to use the "*chaki taclla*", the Andean foot plow, due to their child bearing roles. Local tradition also asserts that a woman should not enter a potato field when menstruating, because she could cause the crop to develop late blight disease. Although these beliefs are being challenged today by many young women who do use the *chaki taclla*, and frequently perform agricultural activities, it is still a common belief that women are less skilled and not able to perform farming activities at the same level as men. This shapes the way women farmers access labor. Both men and women participate in the labor reciprocity system called *huaypo*, but when women reciprocate for work done by men, they are expected to perform activities that are appropriate for their gender, which can make it more difficult for women to get men to reciprocate on their farms.

On the other hand, women's roles as mothers and care givers have been associated with seed management practices. In the Central Andes, potato is considered a living being that needs to be raised and cared for by farmers. Seed potatoes are like children and women are traditionally in charge of selection, storage and management of potato diversity. Tradition claims that men should not handle potatoes in storage because they may damage them.

In order to foster potato diversity conservation, external actors have promoted in-situ conservation and supported custodian farmers to participate in seed fairs and increase the number of varieties they conserve. However, due to gender norms women have traditionally enjoyed less visibility and mobility than men, especially women with young children, which restricted them from participating as diversity custodians, occupying public spaces and receiving full recognition from their deep knowledge of native potato diversity management and conservation. More men than women have come to occupy the role of potato custodians due to their socially recognized participation in public spaces.

A qualitative study conducted with members of the Association of Native Potato Custodians (AGUAPAN) revealed that efforts to strengthen *in situ* conservation have contributed to empower custodians. Yet, ensuring that empowerment processes are gender inclusive remains a challenge. While one third of the members of AGUAPAN are women, they clearly experience difficulty to fully participate in meetings and events due to their numerous responsibilities at home, so they delegate representation to their husbands. Women who reported more participation were often single mothers or older women with adult children.

Based on a qualitative study conducted in the highlands of Peru (Molina et al. 2022).

relations structure the interaction between actors (Wendt 1987). Structure and agency are historically interdependent opposing forces that jointly produce social outcomes (Akram 2010). Thus structure and agency need to be analysed as interacting social fields including a wider perspective of social phenomena incorporating history (Sewell 1992), culture (Archer 2005), consciousness (Elder-Vass 2006; Akram 2010), reflexivity and intentionality (Akram 2010).

Structure refers to social factors including: social arrangements, social relations and social practices which exercise power and coercion in the lives of individuals (Musolf 2003). Initially originating from collective habits, structure finds expression in definite forms such as legal rules, organizational frameworks, moral obligations, popular proverbs and social conventions (Durkheim 1964); it organizes social positions hierarchically where power emanates from those who own the means. Some factors that make up the structural dimension of social life are race, class, sex, ideology, institutions, organizational hierarchy, groups, geographical location, period of history, mode of production, generation cohort, family culture, roles and rules (Musolf 2003). Ultimately, structure can also be divided into three sub sections (Durkheim 1964; López and Scott 2009) connected to the different spheres of life (Fig. 7.2) and Case 7.4.

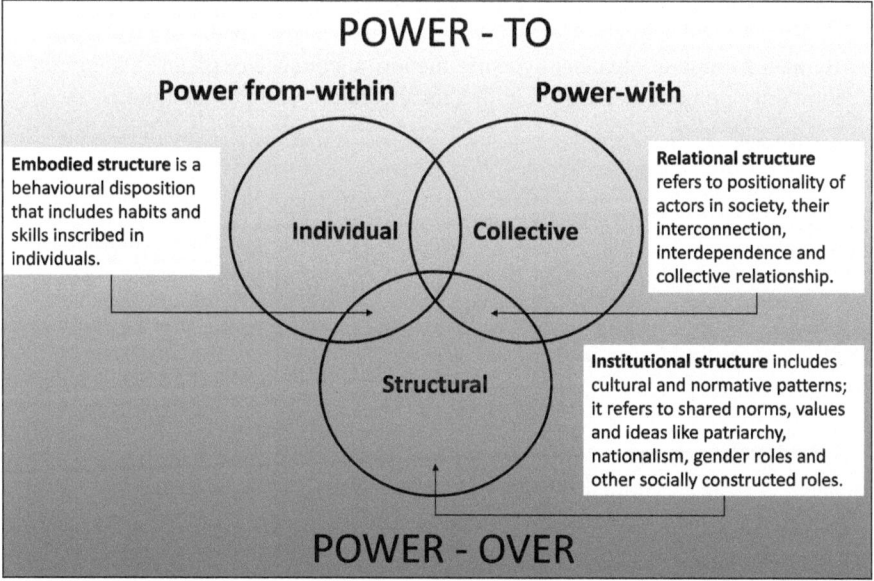

Fig. 7.2 Types of power and their relationship with sub-sections of structure. (Source: Adapted from Polar (2013))

Case 7.4 Pachamama, from Deity to Servant: The Evolution of Gender in Agriculture in the High Andes

Modern gender constructs do not necessarily reflect past ones. Archeological evidence shows that early females in the Andes were big-game hunters (Haas et al. 2020). There is also evidence of the recognition of female supernatural beings among the earliest cultures and first civilizations of the Andes prior to the rise and expansion of the Inca empire. Early cultures such as the Chavín and Yaya Mama conceived supernatural power as both male and female, and work patterns in everyday life appeared to have been mostly egalitarian with indication of women also holding institutionalized roles of authority (Kellogg 2005). Early chronicles of the Inca and Post Inca periods register multiple expressions of deification of feminine sculptures in gold (di Salvia 2013). Yet, the most outstanding representation of the feminine is "Pachamama" or mother earth believed to be the mother of all things, the representation of an animated natural world, and with whom Andean people are in continuous dialogue through tributes and rituals (Pineda 2018).

While Pachamama has been part of the culture and belief system in the Andes since the first recorded chronicles, its representation, meaning and importance in everyday life has evolved through incoming concepts from Christianity and most recently from the green revolution development approach. An example of this evolving process are the comments from Maria (fictional name, approximately 45 years old) farmer and "Mama T'alla" (female community leader) to justify the low attendance of people an agricultural development that started with a local ceremony to Pachamama.

> We used to be united in this community. Always men and women would work together and be leaders. Women select the seed in the house, we separate what is for food and for planting. Mama T'allas were consulted about seed and where to plant. We also read the signals that Pachamama gives us in the local indicators.[1] Now you see our payment to Pachamama is reduced to spilling some alcohol and sharing some coca leaves. Only this is not enough, we need to listen. We are forgetting how to listen to Pachamama because some people say this is not real. Some Christian churches came, they told community members that we should not thank Pachamama, that coca leaves are bad. People have started to skip community meetings; they don't want to share coca anymore. They have told people that men should decide in a family. Now there are fewer people coming to decide together.

Juana (fictional name—approximately 35 years old), a farmer producer of Andean grains, comments in relation to her efforts to produce and participate in organic certification.

> Now people don't produce the same as before. Now when we go to the community fair there are many outsiders. Strange men come to sell products that we don't know, and they convince men that it is good to produce more and more... always more, and to have fewer insects. Pachamama is angry, this is why she sends more insects and production is lower. Now she doesn't speak to us anymore because we are not respecting her.[2]

Based on participant observation in rituals to Pachamama in multiple communities in Pacajes (Bolivia, 2015–2017).

Religion—more specifically, religious beliefs—is one of the blind spots of gender that the international political and academic communities appear to be scared to open up in the context of agriculture and development (Rao and Cagna 2018). Religion is a structure in itself that permeates all spheres of life. As is presented in Case 7.4, the "Mama T'alla" is a leadership position based on the duality of the Andean world view. It is a right to lead conferred to women in specific domains such as seed management in the household (as reflected in Case 7.3). But the leadership position of the Mama T'alla extends to the collective as they negotiate geographical planting distribution and selection of crop rotation among other details of agricultural practice. The Andean belief system linked to Pachamama also created shared norms and values and an "institutionalized" dialogue with mother earth through the language of indigenous knowledge about indicators to forecast the weather.

The superposition of a new religious system happened over centuries and decades after the Spanish conquest, but in Aymara communities of Bolivia and Peru one can still observe the syncretic evolution. Tribute is still paid to Pachamama but her role shifts from deity and mother to a servant of productivity. The original practice of drinking and pouring on the planting plots fermented maize drinks—rich in bacteria that produce bioactive components with enhanced health benefits for humans (Meena et al. 2022), and nitrogen fixing bacteria able to foster plant growth promotion by a variety of mechanisms (Reis and dos Santos Teixeira 2015), has been discontinued. It has now been replaced by pouring distilled cane alcohol (as mentioned in Case 7.4) or drinking and pouring any beverage including industrial beer, soft drinks and tea (Pineda 2018). Basic elements of the practice remain but its reciprocal and regenerative nature has been lost with the incorporation of a new religious structure that relates deity with power, hierarchy, worship and patriarchy.

In the same way women's roles have gradually shifted from complementarity to dependence particularly in collective spaces of the agricultural domain. Duality in representation and dialogue has been replaced by men speaking to men, as mentioned by Juana. As Case 7.4 presents, in regulating women's roles, responsibilities and hierarchical positioning, religion embodies a structure of power that governs individual behavior, regulates relational structure and provides a framework for evolving institutional structures.

Agency refers to the capacity of an agent (person or collective) to act through the independent exercise of their own power based on the meaning they assign to objects and events (Musolf 2003). While some sociological traditions frame agency on the premise of reflexivity from the agent and full consciousness (Archer 2005; Hay 2002; Andersen 2009), others argue for an unconscious component that reflects the effect of structure over the actions of agents (Akram 2010; Elder-Vass 2006). The exercise of agency can both perpetuate or challenge structures of power.

---

[1] Refers to local indigenous knowledge about forecasting weather, by interpreting the signs of nature such as animal behavior, plant distribution, the appearence of stars, wind and other elements (Choquetopa Rodríguez 2021).

[2] In reference to climate change that is generating changes in the manifestation of local natural indicators, making them harder to interpret over time.

Case 7.5 shows how two women leaders assign different meaning to somewhat similar events and how contextual factors influence their exercise of agency to challenge structures. The story of Sara exemplifies agency and full consciousness about her representation role. The experience from Luisa on the other hand exemplifies the unconscious component of structure limiting her participation in the exchange visit. Her concern of what people would say, how others would perceive her travelling alone with strangers, and being absent from her household responsibilities is much too strong. In Luisa's experience, power exercised through gender norms blocks the exercise of agency and perpetuates the structure of power by replication. Sara on the other did not even think about rejecting the invitation. She and other people in the family had heard about Bolivian female migrant workers in Europe. This precedent and her continued work with female agricultural service providers could have acted as a catalyst to enable her exercise of agency.

Case 7.5 Two Women Leaders: Two Distinct Outcomes

Luisa (pseudonym—approximately 57 years old) is a community leader working alongside many other women developing and adapting agricultural innovations to the local context. She tapped into indigenous knowledge to foster the utilization of Andean grains for home consumption when plants were at various stages of development. As a leader of a group of women she worked along side nutritionists and agronomists conducting research and prompting the consumption of leaves and seedheads of Andean grain species, as they matured up to milky stage of grain formation. This enabled farmers to secure nutritious and iron rich food during periods of food scarcity before harvest. Luisa won the first prize in a national agricultural innovation contest, a prize consisting of a fully paid visit to meet and exchange experiences with other farmer innovators in another Andean country. While initially very happy at being recognized publicly as the best innovator, when she understood the full extent of the prize she commented: *"I should have won the second prize, not the first... I will not be able to travel, I have too many responsibilities and what are people going to say? The second prize was a backpack sprayer, that would have been better for me so I can use it for my organic liquid fertilizers. Even some tools for the field or the kitchen would have been better."*

A few days before the trip when tickets had already been purchased, Luisa communicated that she would not travel and seemed discouraged from further participating in other events.

Sara (pseudonym -approximately 35), a community leader and a custodian of agro-biodiversity of Andean grains was globally recognized for her dedication to conserving traditional varieties and local knowledge. She was invited to participate in a global event to share her experience in Rome, Italy. Upon learning that she would be travelling, she happily accepted, packed her bag with her best and most colorful traditional clothing and embarked in a once-in-a-lifetime adventure. Dressed up for the occasion she appeared in a conference

room full of foreign strangers and shared her story in Spanish with simultaneous translation to English and web broadcasting to the world. Her colorful
"cholita" clothing (with its sweeping skirt, embroidered blouse and round felt
hat) raised attention everywhere, with people asking to snap selfies with her,
even on sidewalks outside the event venue. When asked about her perception
of the event she mentioned: "I should have brought my other outfit too, to use
in the evening dinner. I'm an ambassador of quinoa and of my culture. This I
will tell women in the community. People don't always value what we do as
women planting and storing and managing so many seeds of quinoa and also
other plants. They think we are just playing around, and that we should produce all the same variety for the market, but we know the different seeds will
help us in hard times. Our mix of varieties always produces, even in the worst
years you have something. Now I know others see this too".

With her newly strengthened conviction of the importance of agro-
biodiversity fairs, Sara in coordination with other community custodians and
agricultural service providers lobbied with her municipal government to institutionalize agro-biodiversity fairs. Moreover, the gender-related prescience
concerning the importance of plant varietal diversity and conservation under
conditions of increased climate change is now evident globally.

Based on participant observation in capacity building activities for organic
production of Andean grains in Pacajes and conservation of Andean grains
diversity in Cachilaya (Bolivia, 2015–2017).

As the multiple cases presented above depict, both agency and structure and their
observable manifestation connected to power need to be considered to build a holistic conceptualization of empowerment.

7.3   The Concept of Empowerment

The concept of empowerment emerged with feminist movements and since then has
been associated with a wide range of disciplines. While recognizing that there is a
form of power-over, empowerment theorists choose to focus on women's power to
transform themselves, others and the world (Allen 1998; Held 1993; Hoagland
1988). This conception of power-to and its complementary forms of power-with and
power from within frame empowerment as a complex, multidimensional process
that operates both at the individual and collective levels.

In the development arena, empowerment is associated with an alternative perception of development, one that recognizes poverty as disempowerment, and
empowerment as the process that reduces inequalities (Friedmann 1992).

Empowerment implies a holistic understanding of context and environment (Esquivel 2016; Titi and Singh 1995), it implies the expansion of assets and capabilities of poor people (Narayan-Parker 2005). Under this perspective empowerment is dependent on two variables: agency as the ability to make meaningful choices; and opportunity as the aspects of context that affect the ability to transform agency into effective action (Eyben 2005). Thus, empowerment is mainly about agency and it relates to the way development agencies have used the term empowerment over time (Narayan et al. 2000; Narayan-Parker 2002,2005).

This practical conception of empowerment in the development sector builds fully on the philosophical perspectives of empowerment theorists and their definitions of power, ignoring the conceptualization of domination theorists. One sidedness in the conception of power is problematic for it obscures forms of oppression that are intertwined with subordination (Allen 1998; Spelman 1990). Power relations are complex and multifarious. An actor can be both dominated and empowered at the same time and in the context of the same norm, institution or practice (Allen 1998). A clear example of this is presented by Galiè and Farnworth in the analysis of local understandings of empowerment and the way empowerment-related experiences are lived by women and men in agricultural communities in Syria, Kenya and Tanzania. The study coins the concept of "power through" to depict how an individual's ability to exercise agency or not, is connected to processes beyond their control (Galiè and Farnworth 2019).

While the dichotomy of power-over conceived from the perspective of domination vs power-to conceived as empowerment has been debated and questioned by feminist philosophers more than two decades ago (Allen 1998), this debate has not affected the concept of empowerment used in the development sector. While the concept of empowerment has been useful to introduce issues of rights onto the international agenda, it remains a buzzword devoid of its political and transformative essence (Cornwall and Rivas 2015). One clear example is the Sustainable Development Goals agenda 2030 that focuses strongly on gross domestic product growth, conflating economic growth with social progress (Adams and Tobin 2014). The only mention of power in Agenda 2030 is a mention of disparities in wealth, opportunity and power, without any analysis of macro and micro relations that leverage the persistence of such disparities, nor any follow up actions to address them (Esquivel 2016). In contrast, the word empowerment is written into Agenda 2030, mostly in reference to women's and girls' empowerment, yet the term is loosely defined and aligned with an apolitical usage (Esquivel 2016).

The elements of power-over are not addressed in practical empowerment interventions or measurement approaches, despite the fact that in developing countries power has been exercised for centuries through colonialism, patriarchy, capitalism, religion beliefs and cultural practices, and is still being exercised as power-over. This exercise of power through time has created formal and informal structures (Weber et al. 1979), that cyclically legitimize the exercise of power (Martin 1971). In time it has even created psychological barriers (Sampson 1965) that work adversely, disempowering people, robbing them of their self-esteem and individual sense of potency. In this context development that seeks to truly achieve sustainable development must tackle empowerment holistically encompassing: a) the multiple definitions and manifestations of power; b) agency and structure as driving forces of social change; and c) the spheres of life or spaces where power manifests itself (see Figs. 7.1 and 7.2).

7.4 Empowerment in Agricultural Research for Development: To Measure or Not to Measure? that Is the Question!

In the field of agricultural research for development (AR4D) the concept of empowerment is recognized essentially for its instrumental value. While empowerment is generally considered a process, it can also be perceived as an outcome (Carr 2003), with different experiences in the agricultural sector highlighting its contribution to desirable outcomes in nutrition and health, productivity, and management of resources (Elias et al. 2021).

The largest proportion of literature on empowerment in the agricultural sector is devoted to assessment. Assessment of projects, interventions and processes and their effect on empowerment can play an important role in advancing empowerment in at least four ways (Elias et al. 2021): (a) support the design of holistic interventions and policies; (b) monitor the positive or negative effects of interventions on empowerment; (c) build accountability and credibility of interventions; and (d) to use participatory assessment to challenge power relations. However, assessment is not necessarily empowering or desirable. Measurement itself holds an embedded bias aligned to the purpose of measurement, actors involved, knowledge system and methods used.

In a mapping of methods used to assess empowerment in AR4D, 15 different methods were found, most of them emphasizing on assessing agency at personal and relational levels (Elias et al. 2021). Most of the emphasis on measurement is dedicated to show results and outcomes of short- and medium-term interventions to donors and other development actors. This premise has shaped the way measurement is conducted, creating bias by privileging the exploration of agency and shaping the definition of indicators important for external actors who are the final users of the information. Research and development paradigms tend to favor quantifiable knowledge (Nazneen et al. 2014) specially in the agricultural sector. In this scenario the main issue is not about how we enhance or improve assessment, but actually how are all these assessments shaping the way the agricultural sector evolves to foster women's empowerment and truly enhance livelihoods.

In order to re-shape the development agenda to address the multiple manifestations of power that influence women's empowerment, we have to ask different questions during assessment. Perhaps questions should not be about what has changed and the levels of participation, decision or income are, but about what has prevented decisions, income or participation from full realization. These are questions that are unlikely to emerge in a detailed questionnaire or even key informant interviews. Answering this questions may require more in-depth analysis with and by the main actors experiencing the limitations and barriers that thwart empowerment and development.

7.5 Toward a Feminist and Transformative Conceptualization and Operationalization of Empowerment in agriculture

Out of the theoretical analysis presented above and the examples explored in the cases, three main conclusions can be drawn:

Patriarchy, colonialism and religion are strongholds of power that need to be properly disentangled in their relation to gender, class, ethnicity and language, as a step towards addressing empowerment.
Specific power dynamics need to be called by their name and looked in the eye. Many invisible forms of power-over come from patriarchy, colonialism, religion and politics. Avoiding confrontation with such forms of power creates a block that prevents true progress towards empowerment. We must ask how do local women of specific ethnic groups conceive empowerment and agricultural development in their own context? What are their limitations, aspirations and achievements and how do these differ from the ethnocentric perceptions of development agents?

Power is in everything, everywhere and manifests in multiple overt and covert forms facilitating or limiting empowerment.
Empowerment is not something that can be conferred or bestowed on people. It is not an ultimate goal. Empowerment is an intermediate outcome towards wellbeing that requires actors to navigate the ocean of power dynamics. There is no shortcut and there is no area of power that can be avoided in the journey. The empowerment process will necessarily involve a re-negotiation of roles, decisions and spaces across multiple actors. For example, donors and policy makers should be prepared to re-negotiate their development agenda. Researchers must be open to alternative research pathways and knowledge construction processes.

Agricultural development, in its multiple forms is not an ultimate goal but an intermediate outcome towards wellbeing.
For agricultural development to transform into wellbeing for women and other segments of the population, multiple types of innovation need to take place. It is not only about the technology, the productivity, or the income. It is mostly about how these outcomes ultimately translate into positive life experiences and wellbeing. An evolving agenda or agricultural research must be coupled with an agenda of social innovation to address different manifestations of power that prevent women and specific social groups from fully benefiting from agricultural development processes which lead towards broader societal wellbeing.

References

Adams B, Tobin K (2014) Confronting Development: A Critical Assessment of the UN's Sustainable Development Goals. Rosa Luxemburg Stiftung, New York

Allen A (1998) Rethinking power. Hypatia 13:21–40. https://doi.org/10.1111/j.1527-2001.1998.tb01350.x

Andersen H (2009) The Causal Structure of Conscious Agency. University of Pittsburgh, Doctor of Philosophy

Apffel-Marglin F, Sanchez L (2002) Developmentalist feminism and neocolonialism in Andean Communities. In: Saunders K (ed) Feminist post-development thought: rethinking modernity, post-colonialism & representation, Zed books on women and development. Zed, London, pp 159–179

Akram S (2010) Re-conceptualising the concept of agency in the structure and agency dialectic: habitus and the unconscious (d_ph). University of Birmingham

Archer MS (2005). Structure, culture and Agency. In: JACOBS, M. D., Hanrahan, N. W. & Blackwell Publishing, L. (eds.) The Blackwell companion to the sociology of culture.

Bachrach P, Baratz MS (1962) Two faces of power. Am Polit Sci Rev 56:947–952. https://doi.org/10.2307/1952796

Caputi M (2013) Feminism and power : the need for critical theory. Lexington Books, Lanham

Carr ES (2003) Rethinking empowerment theory using a feminist lens: the importance of process. Affilia 18:8–20. https://doi.org/10.1177/0886109902239092

Choquetopa Rodríguez B (2021) Indicadores naturales para pronosticar el tiempo en el Sur de Oruro, Bolivia. In: Bentley J (ed) Agro-Insight and the McKnight Foundation, Cochabamba

Clegg S (1989) Frameworks of power. Sage Publications, London

Cornwall A, Rivas A-M (2015) From 'gender equality and 'women's empowerment' to global justice: reclaiming a transformative agenda for gender and development. Third World Q 36:396–415. https://doi.org/10.1080/01436597.2015.1013341

Cudd AE (2006) Analyzing oppression, studies in feminist philosophy. Oxford University Press, New York

di Salvia D (2013) La Pachamama en la época incaica y post-incaica: una visión andina a partir de las crónicas peruanas coloniales (siglos XVI-XVII). Rev Esp Antropol Am 43:89–110. https://doi.org/10.5209/rev_REAA.2013.v43.n1.42302

Dahl RA (1971) Who governs ? : democracy and power in an american city. Yale University Press, New Haven/London

Durkheim É (1964) The rules of sociological method, etc. Collier-Macmillan/Free Press of Glencoe, New York/London

Elder-Vass DJ (2006) The teory of emergence, social structure, and human agency. Doctor of Philosophy, Birkbeck College

Elias M, Cole SM, Quisumbing A, Valencia AMP, Meinzen-Dick R, Twyman J (2021) Assessing women's empowerment in agricultural research. In: Pyburn R, van Eerdewijk A (eds) Advancing gender equality through agricultural and environmental research: past, present and future. International Food Policy Research Institute, Washington, DC, pp 329–362

Esquivel V (2016) Power and the sustainable development goals: a feminist analysis. Gend Dev 24:9–23. https://doi.org/10.1080/13552074.2016.1147872

Eyben R (2005) Linking power and poverty reduction. In: Alsop R (ed) Power, rights and poverty: concepts and connections. The World Bank, Washington, D.C, pp 15–28

Fernandez J, Polar V, Fuentes W, Ashby J, Villarroel T, Paz R (2012) Caso de implementación SEP 3. Mejorando la calidad de vida de los productores de maíz y maní en el chaco boliviano. In: Thiele G, Quirós CA, Ashby JA, Hareau G, Rotondo E, López G, Paz Ybarnegaray R, Oros R, Arévalo D, Bentley JW (eds) Metodos participativos para la inclusion de los pequenos productores rurales en la innovacion agropecuaria: Experiencias y alcances en la region andina 2007–2010. Centro Internacional de la Papa, pp 57–66. https://doi.org/10.4160/978-92-9060-417-4

Follet MP (2003) Power. In: Thompson K, Follett MP, Urwick LF, Metcalf HC (eds) Dynamic administration: the collected papers of Mary Parker Follett, early sociology of management and organizations. Routledge, London, pp 72–95

Foucault M, Faubion JD (2002) Power, essential works of Foucault, 1954–1984. Penguin, London

Friedmann J (1992) Empowerment : the politics of alternative development. Blackwell Publishers Inc., Malden

Galiè A, Farnworth CR (2019) Power through: A new concept in the empowerment discourse. Glob Food Secur 21:13–17. https://doi.org/10.1016/j.gfs.2019.07.001

Gammage S, Kabeer N, van der Meulen Rodgers Y (2016) Voice and agency: where are we now? Fem Econ 22:1–29. https://doi.org/10.1080/13545701.2015.1101308

Gaventa J (1980) Power and powerlessness: quiescence and rebellion in an Appalachian valley. Clarendon Press, Oxford

Gaventa J (2006) Finding the spaces for change: a power analysis. IDS Bull 37:23–33. https://doi.org/10.1111/j.1759-5436.2006.tb00320.x

Giddens A (1984) The constitution of society outline of the theory of structuration [WWW Document]

Haas R, Watson J, Buonasera T, Southon J, Chen JC, Noe S, Smith K, Viviano Llave C, Eerkens J, Parker G (2020) Female hunters of the early Americas. Sci Adv 6:eabd0310. https://doi.org/10.1126/sciadv.abd0310

Hains RC (2009) Power feminism, mediated: girl power and the commercial politics of change. Womens Stud Commun 32:89–113. https://doi.org/10.1080/07491409.2009.10162382

Hartsock NCM (1985) Money, sex, and power: toward a feminist historical materialism, Northeastern series in feminist theory. Northeastern University Press, Boston

Hay C (2002) Political analysis. New York, Palgrave, Houndmills, Basingstoke, Hampshire

Held V (1993) Feminist morality: transforming culture, society, and politics, women in culture and society. University of Chicago Press, Chicago

Hoagland SL (1988) Lesbian ethics: toward new value, 1st edn. Institute of Lesbian Studies, Palo Alto

Hobbes T, Tuck R (1996) Leviathan, Rev. student ed. Cambridge texts in the history of political thought. Cambridge University Press, Cambridge

Kellogg S (2005) Weaving the past: a history of Latin America's Indigenous women from the prehispanic period to the present. Oxford University Press, New York

López J, Scott J (2009) Social structure. Open University Press, Buckingham

Lugones M (2007) Heterosexualism and the colonial/modern gender system. Hypatia 22:186–219. https://doi.org/10.1111/j.1527-2001.2007.tb01156.x

Lugones M (2010) Toward a decolonial feminism. Hypatia 25:742–759. https://doi.org/10.1111/j.1527-2001.2010.01137.x

Lukes S (2005) Power: a radical view, 2nd edn. Palgrave Macmillan, New York

Machiavelli N, Bull G, Grafton A (2006) The prince, new edition, anniversary edition. Penguin classics. Penguin Books, London

Martin R (1971) The concept of power: a critical defence. Br J Sociol 22:240–256. https://doi.org/10.2307/588888

Meena KK, Taneja NK, Jain D, Ojha A, Saravanan CS, Mudgil D (2022) Bioactive components and health benefits of maize-based fermented foods: a review. Biointerface Res Appl Chem 13:338. https://doi.org/10.33263/BRIAC134.338

Molina CA, Dudenhoefer D, Polar V, Scurrah M, Ccanto RC, Heider B (2022) Gender roles and native potato diversity management in highland communities of Peru. Sustain For 14:3455. https://doi.org/10.3390/su14063455

Musolf GR (2003) Social structure, human agency, and social policy. Int J Sociol Soc Policy. https://doi.org/10.1108/01443330310790570

Narayan D, Chambers R, Shah MK, Petesch P (2000) Voices of the poor: crying out for change. Oxford University Press for the World Bank, New York. https://doi.org/10.1596/0-1952-1602-4

Narayan-Parker D (2002) Empowerment and poverty reduction. The World Bank. https://doi.org/10.1596/0-8213-5166-4

Narayan-Parker D (2005) Measuring empowerment. The World Bank. https://doi.org/10.1596/0 -8213-6057-4

Nazneen S, Darkwah A, Sultan M (2014) Researching women's empowerment: reflections on methodology by southern feminists. Womens Stud Int Forum 45:55–62. https://doi.org/10.1016/j. wsif.2014.03.003

Nelson N, Wright S (2001) Power and participatory development: theory and practice. ITDG Publishing, London

Parsons T (1963) On the concept of political power. Proc Am Philos Soc 107:232–262

Pineda GV (2018) Rescatando a la Pachamama

Polar V (2013) Participation for empowerment: an analysis of agricultural innovation in two contrasting settings of Bolivia (phd). London, SOAS, University of London

Polar V, Babini C, Flores P (2015) Technology for men and women: Recommendations to reinforce gender mainstreaming in agricultural technology innovation processes for food security. Centro Internacional de la Papa. https://doi.org/10.4160/9789290604655

Quijano A (2019) Poder, prácticas sociales y ámbitos de práctica social. Debates En Sociol:151–163. https://doi.org/10.18800/debatesensociologia.201902.009

Reis VM, dos Santos Teixeira KR (2015) Nitrogen fixing bacteria in the family Acetobacteraceae and their role in agriculture. J Basic Microbiol 55:931–949. https://doi.org/10.1002/jobm.201400898

Rowlands J (1997) Questioning empowerment working with women in Honduras. Oxfam Publishing, Oxford

Rao N, Cagna P (2018) Feminist mobilization, claims making and policy change: an introduction. Development and Change 49(3):708–713. https://doi.org/10.1111/dech.12393

Sampson RV (1965) Equality and power, Heinemann books on sociology. Heinemann London, London

Sen A (1995) Gender inequality and theories of justice. In: Nussbaum MC, Glover J, Oxford University Press (eds) Women, culture, and development: a study of human capabilities. Oxford University Press, Oxford

Sewell WH (1992) A theory of structure: duality, agency and transformation. Am J Sociol 98:1–29

Spelman EV (1990) Inessential woman: problems of exclusion in feminist thought. Women's Press, London

Spivak GC (1988) Can the subaltern speak? In: Nelson C, Grossberg L (eds) Marxism and the interpretation of culture. University of Illinois Press, Urbana, pp 271–313

Titi V, Singh NC (1995) Empowerment for sustainable development: an overview. In: Singh NC, Titi V, International Institute for Sustainable Development (eds) Empowerment for sustainable development: toward operational strategies. Fernwood Pub/Zed Books, Halifax/London

Vavrus MD (2002) Postfeminist news: political women in media culture. State University of New York Press, Albany

Wanderley F (2010) La participación política de las mujeres y la agenda de equidad de género en Bolivia. Tinkazos 13:09–31

Weber M, Roth G, Wittich C (1979) Economy and society: an outline of interpretive sociology. University of California Press, Berkeley

Wolf N (1994) Fire with fire: the new female power and how to use it, First Ballantine books edition. Fawcett Columbine, New York

Wendt AE (1987) The agent-structure problem in International relations theory. International Organization 41:335–370

Yanagisako S, Delaney C (1995) Naturalizing power. In: Yanagisako S, Delaney C (eds) Naturalizing power: essays in feminist cultural analysis. Routledge, New York, p 24. https://doi.org/10.4324/9781315021676

Chapter 8
The Tyranny of Tools: The Politics of Knowledge Production in Gender Research

Beth Cullen, Nicole Lefore, Liza Debevec, and Katherine A. Snyder ⓘ

Abstract This chapter examines the trajectory of analytical frameworks and gender tools intended to understand and address the challenges and inequities that shape women's engagement in agriculture. We argue that while a focus on tools in many agricultural development projects can help to identify barriers faced by women, it often does little to address the structural inequality in which women are embedded. We highlight the tendencies of tool-led gender analysis within agricultural projects to: (1) detach tools from their theoretical frameworks, (2) ignore the structural and socio-political obstacles to gender equality in specific contexts, and (3) view tools as silver bullets to address "gender problems" while primarily serving technical agendas. We argue that the co-option, sanitization and de-politicization of gender tools is partly the result of social scientists having to fit within institutional systems dominated by certain scientific logics, frameworks, disciplinary orientations, and social norms. We recommend that meaningful attempts to facilitate gender equality and women's empowerment should be based on politically informed, contextualized understandings that are relevant to people's lived realities, rather than concepts, tools, and data that are externally constructed and applied by outsiders to meet normative scientific, donor, and development agendas.

B. Cullen (✉)
Independent Consultant, London, UK

N. Lefore
University of Nebraska, Lincoln, NE, USA
e-mail: nlefore@nebraska.edu

L. Debevec
Independent Consultant, Ljubljana, Slovenia

K. A. Snyder
Stockholm Environment Institute, Townshend, VT, USA
e-mail: Katherine.snyder@sei.org

© The Author(s) 2025
J. Njuki et al. (eds.), *Gender, Power and Politics in Agriculture*,
https://doi.org/10.1007/978-3-031-60986-2_8

8.1 Introduction

In recent decades "gender tools"—or methods to help systematically identify, describe and understand gender differences—have proliferated within the international agricultural development sector. This chapter critically examines the emergence of gender analysis frameworks and tools and traces the processes by which they have been co-opted, adapted and appropriated in agricultural research for development (AR4D). In the process, we argue, many of these tools have become de-politicized and sanitized. Since the 1980s, numerous critiques of gender frameworks and tools have emerged. These perspectives have been useful in identifying the shortcomings of these instruments to effectively measure women's empowerment. Rather than analyzing the effectiveness of specific tools or critiquing their implementation, we explore the range of tools that have been developed over time and how they are situated between instrumentalist approaches, which seek to understand gender dynamics to achieve project goals, and approaches that emphasize more strategic goals through transforming gender relations. We also examine the ways in which the demand for tools has become internalized and institutionalized within international agriculture research organizations. Due to the volume of tools available, we have focused on only a few to illustrate how they reflect changes in thinking and approaches to gender and development.

As women and social scientists[1] who have worked within international agricultural research for development institutions, namely CGIAR centers, our own work has involved, but not been entirely restricted to, gender analysis. Our perspective on gender tools is informed by these experiences. As researchers, we have participated in designing tools, attending workshops to collate and evaluate tools, critiqued tools (publicly and privately), and have implemented tools in a range of projects and contexts. Our focus is mostly on the CGIAR because, as Okali (2012) comments, CG research practice is well documented, but also CG institutions partner with a broad range of actors undertaking agricultural research and so trends within the CG are likely to be applicable more broadly. However, we acknowledge that this focus potentially ignores practices within national agricultural research and extension systems (NARES) or universities.

While we have been active participants in the tool development and implementation process, we have a growing concern about what we call, the tyranny of tools, where the design and implementation of tools seems to be the primary objective. The pressure to produce and use tools often misses the bigger picture of what the tools are meant to address, which is, better understanding of gender dynamics, the design of entry points to empower women in specific contexts and the means to identify what women and men may want to achieve. We argue that a dependency on tools can result in a lack of meaningful engagement with feminist thinking. We also consider the degree to which tools facilitate a "dumbing down" and over-simplification of the contributions that social science can make to agricultural

---

[1] A political scientist and three anthropologists.

development. As a result, the gender analysis that arises from the implementation of tools does little to advance more just, and equal gender relations.

Tools themselves are not to blame for this situation. Indeed, tools have been useful for expanding our knowledge of gendered livelihoods. However, while tools make important contributions, how they are perceived and used needs rethinking, and this requires greater acknowledgement of the limits that tools have for achieving social change. Our "theory of change" (Vogel 2012) argues that rather than designing more tools, or tweaking the ones that exist, transformational gender research requires more critical interrogation of research processes and objectives, as well as the contextual factors that influence tool design. This involves: holistic analysis of systemic structural and institutional barriers (van der Burg 2020); attention to politics and power and their complexity (Leach et al. 2020); awareness of the legacies of imperialism (including patriarchal power structures) (Cummings et al. 2022); challenging existing organizational arrangements, particularly how research questions and solutions are defined, allocation of authority and resources, and cultural and scientific norms (German et al. 2010).

8.2   Conceptual Framework and Methodology

Using historical tracing, this chapter looks at the origins of tools and the ideas that underpin them, the discourses and agendas surrounding them, the actors involved in developing, promoting and using them, and the institutional and organizational contexts that shape them. "By tracing the emergence, development and expansion of instruments over time with a view to the concrete practices that articulate and sustain them, we find that such tools have their own social histories and trajectories" (Simons and Vobs 2018, p. 15–16). Our approach is informed by emerging trends in feminist and agricultural research, and the so called "relational turn" in the humanities and social sciences (Selg and Ventsel 2020). From this perspective, the world consists not of discrete, individual "things" but of relations. Relations are co-constitutive, i.e., that entities, like gender analysis tools, cannot be understood in isolation from their social context or from the relations that give rise to them. This position therefore encourages systemic and relational ways of thinking, holistic accounts and a recognition of situated and diverse knowledges.

Our analytical approach is also informed by the field of science and technology studies (STS), which investigates the social configuration of scientific practice by illustrating how scientific knowledge production and technology development are linked to society more broadly (Jasanoff et al. 1995. STS approaches show how scientific knowledge and technical interventions are also social constructions that are embedded in society. Thus, in order to understand processes of knowledge production and technology development, it is necessary to analyze the social structures that co-produce them. This leads to a focus on scientific disciplines, research institutes, and their structures and cultures, and the power relations that permeate these arrangements (Haraway 1991; Harding 1991; Latour 1987; Star 1990; Wajcman

2010). Knowledge production is a social process and research is part of social structures and infused with power relations (Bourdieu 1988), as such it is important to be reflexive about research institutions, cultures and processes. Our chapter explores the institutional contexts, social relations, and research practices surrounding tool development for gender analysis. Our understanding of power is influenced by Foucauldian (Foucault and Rabinow 1984) and post-humanist perspectives (Haraway 1991; Barad 2007; Braidotti 2021) which see power as operating within networks of relations between human and nonhuman actors, which include organizational structures and institutional arrangements, social rules and cultural norms, technologies and tools.

The term "tool" has come to be used generically, often with no clear definition. Because terms are often used interchangeably, we included frameworks, approaches, tools and methods in our analysis. For the purpose of tracing the trajectory of tool design and use, we also use tool as our primary nomenclature. However, it is important to understand the distinction between framework, tool and method. A "framework" can be thought of as a "methodology" which outlines the theoretical approach upon which gender planning, research and analysis is based. In addition to setting out the overall rationale or approach, each framework also consists of a set of "tools" or "methods" that enable planners, practitioners, or researchers to carry out gender analysis. As such "gender analysis" involves both the conceptual frameworks used to orient analysis and planning, as well as a range of methods or tools for the collection of data. Although tools may be regarded as simply representing different ways of collecting and analyzing data to incorporate gender within research and development processes, it is important to recognize that embedded within them are certain epistemological and theoretical positions, and worldviews.

Reviewing the gender and development literature, as well as prominent tool portals, such as the CGIAR Gender Impact Platform, and the CCAFS website, we have identified publicly available frameworks, tools and methods produced between 1985 and 2021. A list of tools is found in Appendix. While this is not an exhaustive list of all existing gender tools, it includes many of the predominant and most frequently used ones that have emerged since the 1980s. To identify these, we searched for tools related to gender, women and agriculture practices and value chains, but excluded tools that were for adjacent subjects such as fisheries, environment, natural resource management.

In this chapter, toolkits were deconstructed to analyze each tool that had been collected under a framework within the toolkit, whereas iterations of a tool were also reviewed individually as a singular tool, if application was adapted for other purposes or sub-sectors, i.e., some tools we included as a single tool have been adapted through multiple iterations for specific value chains, projects or organizations. For our purposes, tools were analyzed to understand the range of aims and approaches over time, rather than to undertake a quantitative assessment. The tools we review are largely produced and used by various key organizations within the agricultural development sector. These include AR4D organizations such as the Consultative Group on International Agricultural Research (CGIAR), global policy groups such as the World Bank and the Food and Agriculture Organization (FAO), major funders such as US Agency for International Development (USAID), and philanthro-capitalist donors such as the Bill and Melinda Gates Foundation (BMGF). These organizations are the main players in the international agricultural development arena, and significantly influence the design and implementation of gender research.

To track the trajectory of tools, we looked at the year the tool was published and widely recognized. We considered the donor that funded tool development and the institution that designed and piloted it. We identified the stated aim of each tool and asked if the tool claimed to be measuring or supporting the efficiency of development interventions (instrumentalist focus) or whether it was oriented towards empowerment objectives. We analyzed the methodological emphasis of each tool, i.e., if the tools were being used to collect data and measure factors, or if they sought to influence or facilitate social change. We analyze tools as cultural products; their content is contingent on their social environment. We also consider how tools might produce, reproduce or reinforce patriarchal, capitalist ideas of social organization, and other socio-political agendas.

8.3   A History of Gender Analysis in International Agricultural Development

Integrating gender into agricultural research has roughly followed the broader trajectory of incorporating gender into international development programs and projects. We outline a brief history of gender inclusion within agricultural development to provide context and to describe the socio-political trends that informed the rise of key tools and frameworks.

The development of tools to incorporate gender can be seen as a reaction to how agricultural development was practiced in the colonial era. As many scholars (Amediume 1987; Oyewumi 1997; Hodgson 2001) have observed, ideologies of gender that permeated colonizing societies came to be imposed in the colonies. For example, agricultural extension directed interventions at men who were believed to be the primary decision-makers and principal farmers. Women were thought to be limited to work within the domestic sphere. Households were also conceived as nuclear, rather than complicated extended family units. In the post-colonial era, development interventions built upon these ethnocentric colonial biases, and contributed to the imposition of a Western, male-biased model that largely bypassed women (Gardner and Lewis 2015). These historical processes served to redefine gender relations and entrench new forms of male dominance (Farhall and Rickards 2021). From the 1970s onwards, a shift in thinking occurred with the 1970 publication of Ester Boserup's book, "Women's Role in Economic Development". It highlighted the critical role of women in local and international economies by focusing on agricultural production in sub-Saharan Africa. This landmark publication prompted increasing interest in the relationship between gender and development.

Feminist scholars from a range of disciplines demonstrated that women play essential roles in the economic, social and cultural development of their societies, and thus are central to development.

In response to Boserup's work, the women in development (WID) paradigm emerged in the 1970s, giving greater visibility to the role of women in agriculture. This paradigm drew upon liberal feminist discourse and aimed to improve women's inclusion in development processes. The WID paradigm often focused on increasing women's participation to make development projects more successful and efficient. WID was closely related to the modernization paradigm that dominated mainstream development thinking from the 1950s into the 1970s. It focused on women's inclusion in processes of "modernization" (Rathgeber 1990), based on an assumption that an increase in women's economic participation would lead to increased equity (Moser 1989). Proponents of this approach therefore argued that it was a more efficient use of resources to include women in development projects (March et al. 1999), though they did not propose the use of tools. Rather, they hoped that demonstrating how investment in women's productivity could result in economic returns as well as social returns, would convince planners to direct development resources to women (Razavi and Miller 1995). While WID did much to highlight the roles and work of women, its instrumentalist approach often focused on ensuring women's practical needs (such as water, food, nutrition) were met, rather than addressing higher strategic needs (such as equitable access to and control over resources). The focus on the practical, over the strategic, led feminist scholars to argue that agendas of efficiency and economic growth do not necessarily align with goals of wider gender equity (Caulkin 2015; Jackson 2007; Chant 2016).

The launch of the United Nations Decade for Women (1975–1985) saw a growing institutional commitment to women's issues. While this was prior to the promotion of tools as such, this shift in emphasis influenced the early movement toward tool development. This was partly prompted by feminist successes in the Global North which enabled women to reach management positions within development organizations and push feminist issues onto the policy agenda. By the early 1980s many development agencies had determined to "do something" for women (Gardner and Lewis 2015, p. 157). This period also saw criticism of the WID paradigm intensify, for example through the women and development (WAD) approach which drew on dependency theory and Marxist-feminist analysis (Rathgeber 1990). Much of this critique was based on post-colonial perspectives which highlighted that WID, and the Northern theories of development on which it was based, were premised on Western ideals that do not translate to Global South contexts.

Feminist scholars in the Global South also demanded that development researchers and practitioners acknowledge the diverse, multi-layered and complex realities of women. There was a growing realization that targeting women in development projects was not sufficient to address gender inequality. This recognition gave rise to the gender and development (GAD) approach in the 1990s. Rather than an exclusive focus on women, GAD called for attention to social relations between men and women, and how these relations are influenced by broader social dimensions (Kabeer 1999). This paradigm emphasizes the structural factors that underpin inequality, such as considering how class and race intersect with gender. This intersectionality results in differential positions of women and men in specific social contexts. Drawing on theoretical work by both Global South and Global North feminists, the GAD approach argued that development interventions cannot succeed if they only address economic issues without considering social and political structures. In contrast with the WID paradigm, GAD argues that the whole model of development needs to be transformed by addressing social and political dimensions as well as economic factors.

Although the language of "gender equality" and "women's empowerment" has become ubiquitous in the international development sector, it has also been accompanied by a policy shift whereby development institutions seek to address poverty through neoliberal models that emphasize and encourage the reach of global capital (Wilson 2015). Such rationales are reflected in the current preoccupation with value chain approaches, market segments and greater economic inclusion. Several scholars have examined how the World Bank, the Nike Foundation and other development and philanthropic agencies have promoted a focus on women and girls as "smart economics" (Hickel 2014, p. 1356). The emphasis that is placed on individual attainment, "freedom" and economic empowerment is deeply embedded in neoliberal and western assumptions (Hickel 2014; Chant 2016; Jackson 2007; Cornwall 2016; Caulkin 2015). Such approaches to gender equity, and the models of development that they promote, are often at odds with, or undermine existing socio-cultural arrangements in the Global South (Jackson 2007).

These conceptual shifts over time, and the socio-political agendas that underpin them, have all contributed to the development and transformation of frameworks and tools for gender analysis. We explore how these ideas play out in relation to specific tools in the sections to follow.

8.4 Review of Dominant Gender Frameworks and Tools

Several significant gender frameworks emerged during the 1980s and 1990s, which aimed to address gender-related concerns and to gather gender disaggregated data for monitoring and reporting. As these frameworks were the first of their kind and paved the way for the proliferation of tools which occurred from the late 1990s onwards, we discuss each of them in turn, and how they correspond to the historical trends outlined above. In recent times, as attention to gender has been mainstreamed in development, the demand for practical instruments has intensified. This focus on tools is based on an assumption that simplified, easy-to-use tools will enable practitioners to systematically incorporate gender into development processes (March et al. 1999, p. 9), while also addressing donor concerns about impact and accountability. As these tools are now numerous, in the latter part of this section we focus only on illustrative examples of recent tools that reflect dominant socio-political concerns.

Tools were initially developed to measure whether or not projects were having a positive impact on women. This interest in impact was shaped by the recognition that many development initiatives had been perceived as failures, with some scholars and development philanthropists linking the failures to the lack of integration of women into project design and activities. Early gender tools were thus designed to measure the impact of donor investments, by measuring changes for women in targeted areas. The Harvard Analytical Framework,[2] published in 1985, was one of the first frameworks targeted at understanding women's economic development (Overholt et al. 1985). Designed by the Harvard Institute for International Development (HIID), and supported by the WID office of USAID, the framework emerged at a time when the "efficiency" approach to integrating women in development was gaining prominence, i.e., that greater economic inclusion would lead to greater gender equity.

Tools were needed to track the outcomes of targeting women, with the economic impact of interventions being the primary concern and increased social equity a secondary concern. The Harvard framework analyzed gendered labor roles, as well as access to and control over assets and income, to assess the impact of technology change and economic development (Okali 2012). Aligned with the WID approach, the framework intended to help planners design more effective and efficient projects by better understanding whether resource allocation to women could improve economic productivity whilst also achieving development goals (March et al. 1999). As a result, the framework tended to treat women separately from men and did not incorporate gendered relations within families or the wider community. This became one of the best-known frameworks in agriculture and rural development and includes a set of data collection tools that have been widely used for sex disaggregation as well as analysis of gender roles, and access and control over assets in farm households (Okali 2012).

As the GAD approach emerged in response to the shortcomings of WID, attention to women within wider social structures emerged. The Moser framework was developed in the early 1980s for the World Bank by Caroline Moser, based at the Development Planning Unit (DPU) at the University of London. It was designed as a gender policy and planning framework rather than a data collection method. Tools outlined within the framework focus on women's triple roles in production, reproduction, and community activities; women's practical and strategic gender needs; and analysis of policy processes. Influenced by GAD, it recognizes the reproductive work carried out by women alongside productive work, making work visible that previously tended to be invisible. It did not totally discard the efficiency approach advocated by WID, but instead made a distinction between practical or basic needs, and strategic needs, i.e., interventions that might alter power relations.

Despite its attention to power, the framework has been criticized for treating women as homogeneous by not accounting for differences based on age, class and ethnicity. Nevertheless, the framework questions the assumption that planning is a

---

[2] Also known as the Gender Roles Framework or Gender Analysis Framework.

purely technical task and explicitly states that "the goal of gender planning is the emancipation of women from their subordination, and their achievement of equality, equity, and empowerment" (Moser 1993, p. 1). It therefore recognizes that gender planning is political and encourages users to question the assumptions underlying development interventions.

Prior analysis has highlighted a dichotomy between tools that emphasize efficiency and those that stress empowerment (March et al. 1999). One of the first tools to emerge with an explicit empowerment agenda was the women's empowerment framework (Longwe 1991), designed by sociologist Sara Hlupekile Longwe, a gender and development consultant in Zambia. In contrast to earlier approaches, and in acknowledgement of the fact that all development interventions are political, Longwe developed a framework that could measure the impact of projects on women's empowerment. It was intended to help development planners and practitioners question what women's empowerment and equality means in practical terms, and to critically assess the extent to which development interventions support this empowerment.

Longwe sought to measure a development intervention's impact as either neutral, positive or negative for women, and the framework did so across five different levels of empowerment: control, participation, conscientization, access and welfare. Her thinking was clearly influenced by Freire (1970) with its inclusion of conscientization and women's perceptions about their own power and place in society, empowerment being something that cannot be granted but must be self-generated. By striving to define empowerment but also address oppression the framework goes well beyond a focus on material wellbeing and resource access that had been the emphasis of previous frameworks. Longwe developed the framework in 1991, and it became UNICEF's official gender policy in 1994. Although her work contributed to the conceptualization and recognition of empowerment for achieving development goals, organizations struggled with how to operationalize and implement the framework.

Alongside mainstreaming approaches adopted by development bureaucracies, this period also saw the emergence of more bottom-up, transformative processes. These include the gender analysis matrix (GAM) which was developed by Rani Parker in 1993 together with NGO practitioners. This community-based technique is structured around two main concepts: analysis at four "levels" of society (women, men, households and community); and analysis of four kinds of "impact" (labor, time, resources and cultural factors) (March et al. 1999). Unlike other frameworks, it is explicitly designed to be used by community members and is intended to start a process whereby gender roles are questioned and challenged, and it is transformative in orientation. This tool is similar in style to a range of participatory methods, including rapid rural appraisal (RRA), participatory rural appraisal (PRA), and appreciative inquiry, which seek to facilitate self-observation and awareness, followed by setting common goals to achieve a shared vision (Chambers 2008). Such techniques therefore seek to identify and analyze gender differences and their implications by carrying out research *with*, rather than *on* the men and women concerned.

Feminist thinking about women's rights, agency and empowerment entered institutional landscapes in the 1990s via gender mainstreaming efforts, but many of the concepts were co-opted and diluted in the process (Cornwall et al. 2007). Critical questioning of what such concepts mean in practice and the extent to which they can be operationalized led to the design of new frameworks. The social relations framework,[3] arose as a critical response to the Moser and Harvard frameworks and sought to provide a more radical and less economistic conceptualization of gender equity. Developed by Naila Kabeer (1999) at the UK-based Institute of Development Studies (IDS), as the name implies, the framework focuses on women's empowerment via a social relations approach. It emphasizes the role of institutions and distinguishes four key institutional sites where gender inequalities are created and reproduced, namely the state, the market, the community, and family/kinship.

Analyzing institutions is of critical importance as it implies that the underlying causes of gender inequality are not confined to the household and family, but rather are systemic and structural. As such, the emphasis is on transforming the social structures, processes and relations that give rise to women's disadvantaged position. This approach is also based on the assumption that development is not simply about economic growth, technical efficiency or improved productivity, but about improving human well-being. As institutions produce, reinforce, and reproduce social differences and inequalities, their policies and practices must be scrutinized to uncover their core values and assumptions. Critically, efforts to address women's subordination must go beyond the reallocation of economic resources and involve a redistribution of power.

The social relations framework used concepts rather than tools to analyze relations between people and "institutions" to give a more holistic and structural analysis of gendered inequalities. However, the demand for tools that could measure and quantify change and empowerment continued, leading to the development of the women's empowerment in agriculture index (WEAI) in 2012 (Alkire et al. 2013). Designed to measure women's inclusion and agency in the agricultural sector, it was funded by the United States Feed the Future Initiative, and developed by researchers at the International Food Policy Institute (IFPRI) and the Oxford Poverty and Human Development Initiative. This index-based survey tool aims to benchmark levels of gender inequality to design appropriate interventions. Designed as a household survey instrument, the tool scores women's level of decision-making and control using ten indicators across five domains: production, resources, income, leadership and time.

The subsequent project-level WEAI (pro-WEAI) comprises 12 indicators organized into three domains: intrinsic agency, instrumental agency and collective agency (Malapit et al. 2019). Both tools draw on Kabeer's empowerment framework (1999, 2005). Although such tools help advance understanding and contribute to data collection, the WEAI has been criticized for relying on categories and codes that result in distorted representations of household relations and that do not account

---

[3] Also known as the Kabeer framework.

for diverse circumstances facing farming households. It also has been critiqued for focusing mostly on individuals rather than the relational dynamics that create inequity (Addison et al. 2002). Perhaps the biggest criticism of the tool is its underlying assumption that it is possible to obtain an accurate measure of women's empowerment using a rapid survey and quantifiable data (Addison et al. 2002). This critique challenges the assumption that focusing on women's individual autonomy, decision-making power and aspects of their lives that are tied to markets, will facilitate economic growth (Cornwall 2018). Although widely used across the sector, the quantitative, survey-based tool sits uneasily alongside feminist critique of quantitative approaches (Farhall and Rickards 2021). This tool has been trialed, refined and revised since its initial launch, leading to an updated version of the original tool, a qualitative companion tool, and new iterations such as the women's empowerment in livestock index (WELI) (Colverson et al. 2020).

Such critiques are not restricted to the WEAI tool. Kabeer (1999) describes how many assumptions embedded in donor and development institutions, and by extension their tools, are rooted in mainstream economic theories of the market. Ashby and Polar (2021) developed the G+ tools for the CGIAR, for example, to direct researchers to analyze women as actors in market segments, i.e., as producers and consumers. The aim of these tools is ultimately to influence crop breeders to better design their varieties to address the needs and constraints of women (Tarjem 2022). While addressing these needs is important for crop breeding, it cannot be assumed that doing so will lead to greater gender equality and women's empowerment. Wider contextual factors influence these outcomes and indeed participation in markets is rarely equal and does not in itself lead to greater equity. There are most often winners and losers in markets, and tools alone cannot address these structural inequities. Thus, we must consider how tools promote normative agendas and notions of economic participation, which may strengthen rather than challenge patriarchal and capitalist institutions (Tarjem et al. 2022).

Our analysis highlights how gender tools have "evolved in tandem with the evolution of 'gender' in development … and based on very different understandings of the nature of power and inequality (and) … differ regarding their assumptions of what needs to be analyzed and addressed" (Warren 2007, p. 190). Such assumptions are evident, for example, in the focus on the household and domestic sphere, which are deeply rooted in the Western concept of the nuclear family and notions of individualism (Jackson 2007). The proliferation of tools suggests that achieving greater equality for women in agriculture is a technical rather than a political process.

However, tools themselves are embedded with political implications and are not simply neutral devices. They contain tacit assumptions, which can often remain hidden, even to the researchers who use them (West and Schill 2022). The assumptions that inform tool development are invariably laden with value-based and political dimensions in their design and use. This is important to acknowledge, particularly in light of the growing recognition that research tools do not just provide descriptions of reality, "but actively help to make and bring these realities into being" (West and Schill 2022, p. 2). As such, tools are potentially powerful actants within AR4D assemblages that can inform the direction that gender analysis takes and the "solutions" that such analysis informs. Recognizing the values and assumptions embedded within tools, as well as how they reflect and reinforce certain objectives, is therefore critical for politically informed gender research.

8.5 Critical Reflections on Tools and Tool Use

The key frameworks outlined above do not capture the diversity of tools that exist, but rather give an overview of the various theoretical and methodological approaches to gender analysis over time. These tools are still used and applied in various ways by different institutions and projects, and in recent years several new tools have been developed to address specific challenges within the agricultural development sector. As well as creating new tools, organizations are also increasingly producing guides and toolkits of existing or adapted tools for gender in agricultural development. Practically, the gender guides and toolkits provide a menu of summaries, purposes and repositories to find existing instruments and methods. For example, a manual developed by CCAFS guides researchers on what data to collect, how to collect it (tools) and also provides a template for reporting data, though it is not always clear to whom data is being reported and why. Another characteristic of the guides is to organize the tools by stage of the project. A few guides recommend a tool for each step and stage of the project cycle, particularly tools aimed at improving gender responsiveness of project designs. Although these toolkits and guides are helpful and informative they fail to address certain fundamental issues surrounding tool use. Across the dozens of tools and toolkits that we reviewed, we observed a number of significant shortcomings.

Firstly, practitioners and researchers often separate tools from their associated frameworks. As is evident from the previous discussion, the frameworks that dominate the agricultural development sector originate from different contexts, often with different ideological or philosophical agendas. Although the terms "framework" and "tool" are often used interchangeably, their meanings are distinct. A framework outlines the theories, concepts or philosophical approach upon which research or planning is based, while tools are components of these frameworks and relate to specific methods that enable planners, practitioners, or researchers to carry out gender analysis. The underlying political rationale of the frameworks within which tools have been designed is often overlooked in tool implementation. If research tools work to produce social realities, the political dimensions of these tools are fundamentally linked to the differing social effects that they help to produce. Therefore, researchers must be conscious of the tools that they select, together with their associated frameworks, and how these may work to strengthen certain realities and weaken others (West and Schill 2022).

A second point is that most tools do not include training on the ideas that led to their development. In the 1990s, gender specialists emphasized the need for targeted and in-depth training on frameworks and tools (Tsikata 2001). Researchers and practitioners both argued that tool users needed to be trained on the theoretical foundation of frameworks and their related tools in order to reach the goals for which tools were designed. Gender analysis must be carried out in a systematic way; adequate time, skill and preparation are essential (March et al. 1999). As Kabeer writes, "each method is only as good as its practitioner" (1995, p. 112). Warren (2007) argues that training which remains restricted to the tool, and the skills needed to apply it, renders the tool ineffective. While the number of tools is increasing, this call for high-quality training continues to be ignored. As Tarjem et al. (2022) found when observing the pilot of the G+ tool, the same tool may be interpreted and used differently, particularly when applied by biophysical scientists who are not trained in gender and not committed to the task.

Another trend is an inclination to "standardize procedures" as well as a "strong preference for the use of checklists, indicators and measuring" (Halsema 2003 p 83). As Farhall and Rickards argue, in "the rush to operationalize a gender mainstreaming approach there was too much of a focus on tools, indices and metrics" (2021, p. 5). This reflects the booming interest in indicators from the mid-1990s onwards, which saw a wide range of efforts to measure social phenomena, partly driven by a demand for metrics from policy makers and donors, and a growing desire for accountability. As Merry (2016) points out, indicators and metrics are appealing because they appear to be objective, scientific and transparent, and claim to stand above politics, offering rational, technical knowledge.

However, this shift has had significant consequences for gender analysis. Toolkits, indicators, metrics and checklists grow, often as part of mainstreaming efforts, while gender analysis has become de-politicized, reduced to a box-ticking activity (Standing 2004). Cornwall et al. note, that "where 'gender' comes to be represented in the guise of approaches, tools, frameworks and mechanisms, these instruments become a substitute for deep changes in objectives and outcomes" (2004, p. 4). In addition, Halsema argues that the standardization of procedures, and the preference for checklists, indicators and measuring, indicates a strong empiricist orientation in gender planning (Halsema 2003, p. 75). This orientation clashes with feminist methodological positions that question the possibility of achieving "objective" and "true" knowledge by following strict scientific procedures to understand complex and messy social realities.

Finally, almost no tools provide guidance on analysis once the data is collected. So, while there is convergence around the importance of collecting gender disaggregated data, there is little discussion in the literature on what should be done with all the data being collected, in terms of how to make data open access, or how to analyze and interpret data. By 2012, gender specialists had already observed the availability of numerous gender tools (Russo 2012), but as IFPRI noted in 2012: "Over the past decade, donor organizations, researchers, and development practitioners have recognized the importance of collecting mixed methods gender and assets data … Nonetheless, many researchers and practitioners remain unsure of why or how to do this" (IFPRI 2012, p. 1).

Our analysis indicates that most gender analysis tools are predominantly data collection instruments producing gender disaggregated data, which is often a donor requirement. Throughout the period under review, quantitative tools for collecting and analyzing data prevail, although there has been a steady call for qualitative instruments. Less than 10 percent of tools identified were for facilitating change, or transforming gender relations, i.e., guidelines and methods for reflective, planning, and behavior-change oriented activities. Such tools often apply participatory methods to raise awareness and understanding of practical constraints, and collectively (in households, groups, cooperatives or communities) reflect on, and plan solutions. Participatory tools are meant to enable those most affected by an issue to voice their perspectives and experiences, whilst also attempting to address power imbalances between researchers and research participants. Nevertheless, overall, the main objective of tools is data collection, often to inform the design and implementation of technical interventions, reflecting institutional and donor concerns.

8.6 Institutional Context and Marginalized Position of Social Scientists

So why are tools continuing to grow and proliferate, whilst simultaneously failing to close "gender gaps" or contribute in a meaningful way to women's empowerment? To answer this question, it is not enough to look at the tools themselves, and the frameworks that guide their use, rather it is necessary to analyze the specific organizational and institutional contexts in which these tools are being developed and applied. This section will focus on analyzing the position and role of social scientists,[4] including gender specialists, within the agricultural research for development sector in order to understand the context in which these dynamics are materializing and what might be contributing to them. Our historical tracing suggests that tools are informed by broader socio-political concerns, and that their development and use is influenced by dominant logics, particularly institutional settings. With this in mind, we argue that the marginalized role of social scientists and gender specialists within AR4D institutions has, in part, contributed to the scientification and instrumentalization of gender analysis and social science more broadly.

Despite many efforts over the years to institutionalize social science and diversify research staff and approaches, as a sector historically dominated by natural sciences, the notion of the "rational scientific objective man" (Bryant and Pini 2006, p. 267) continues to loom large in institutions focused on agricultural research and development. The agricultural sciences are still heavily positivist in orientation, operating on an assumption that reality exists independently from the observer, and that it is possible to uncover objective and accurate knowledge of the world through empirical, scientific methods. This is in marked difference to the constructivist worldview that underpins many of the social sciences, where reality is understood

---

[4] We refer to social science in a broad sense as research that includes sociological or anthropological disciplines, as well as political science, geography, economics and women's or gender studies.

as constructed by the observer, research results are valid only in specific contexts, and therefore multiple perspectives are accepted (Neef and Neurbert 2011). The biological disciplines have historically set the standard for what is deemed to be "real science" in agricultural development research, their approach to knowledge production providing the model that other disciplines must conform to.

As the "newcomer on the block," social sciences are usually placed in a service role and considered to be of secondary importance. Social sciences, including gender analysis, are often considered to be unscientific, their methods too qualitative, subjective, anecdotal, and time-intensive. As Pyburn and van Eerdewijk point out, "dominant notions of what constitutes 'science' fail to recognize and allow space for feminist approaches, frameworks, data, and insights" (Pyburn and van Eerdewijk 2021, p. 51). The organizational dynamics and scientific culture within agricultural development organizations also feed into and reinforce the notion that, for social scientists, "the knowledge they produce is perceived to lack the accuracy, law-like character, value freedom and rigor of 'real science'" (Brinkmann et al. 2020, p. 31). The proliferation of gender tools in the agricultural development sector could be seen as an attempt to scientize social science and thus prove its worth, rigor and validity by developing more quantitative means of assessment.

Unequal disciplinary hierarchies and power dynamics also intersect with gendered aspects of agricultural development organizations. As Acker (1990) points out, organizational structures are not gender neutral, and although this is apparent within the agricultural development sector, it is often ignored in discussions of gender research processes. The agricultural research and extension landscape has historically been male dominated (Farhall and Rickards 2021). This has meant that women have had little influence over the directions that agricultural research has taken. Although this is shifting, the biophysical scientists who command the sector, and who are considered to carry out the primary science, are still mostly men, whereas social scientists, particularly gender experts, tend to be women. For example, analysis of the CGIAR system conducted in 2001 found that 73 percent of its 7851 staff members were men, with the majority of staff concentrated in the agricultural, natural and life sciences (Rathgeber 2006, p. 52).

Due to their disciplinary status and gender, women social scientists are doubly marginalized (Verma et al. 2010, p. 272). In her 1986 report on gender work in the international agricultural research centers (IARCs), Jiggins comments, "It is commonly said that gender issues are relevant to agricultural research only as downstream considerations, that they have little place in discussion of how research is organized or the methodologies that are used" (1986, p. 27). This situation has not significantly changed over the years, reflected in the fact that social scientists are largely relegated to carrying out ex-post analysis, placing them in the position of critics. For example, in crop improvement research, the input of social scientists has been considered most relevant in the aftermath of crop breeding processes, such as adoption studies and impact assessments, and so they had limited influence over priority-setting and varietal design from the outset (Tarjem 2022).

Because of their exclusion from early stages of research design and conceptualization, social scientists are often perceived by their biological colleagues as "vocal anti-change naysayers who preached about how technical people had gone wrong" (Rhoades 2006, p. 409). Rhoades goes on to say that women anthropologists and sociologists have been accused of "carrying the 'gender thing' too far through emotional outbreaks and verbal haranguing of patriarchal male colleagues," feeding into social stereotypes of women as critical nags, further contributing to their marginalization.

The standardization and de-politicization of gender tools is perhaps partly a result of social scientists having to fit within institutional systems that are dominated by certain logics, frameworks, disciplinary orientations and social norms, thereby supporting rather than challenging the status quo. As Cornwall and colleagues argue, "The institutional context of large development bureaucracies not only leads to the simplification of gender and development ideas, it also transforms them," because organizations tend to incorporate information on their own terms, privileging that which fits in with their own views of the world, their analytical frameworks and research orientations (Cornwall et al. 2007, p. 8). Work by Springer (2020), for example, demonstrates that although gender advisors within international agricultural development organizations recognize that qualitative data better captures the lives of the women that they aim to assist, many want *more* quantitative measures due to the highly gendered nature of their occupation and the influence that numbers have within organizations that prioritize indicators and metrics due to their apparent reliability and validity.

This highlights the ways in which gendered organizations and bureaucratic tools intersect, with implications for knowledge production and development outcomes. Tools can therefore be seen as a site of struggle and contention over disciplinary power and gender relations within the sector. Social scientists, usually women, are perhaps appropriating, adapting and developing tools—particularly quantitative tools—as a way to insert themselves into a masculine space, based on an assumption that this will enable them to assert authority and gain disciplinary respect. The fact that the CGIAR Gender Platform, which collates gender analysis tools, was "launched in part to raise 'gender knowledge' from being lowest in the ladder of sciences within the CGIAR" (Resurrección and Elmhirst 2021, p. 171) is indicative of this struggle. Although tool development has been seen as a means to address the lack of common ground and limited integration between biological and social disciplines (Polar et al. 2022), and to manage institutional and interdisciplinary power relations by strengthening the position of gender specialists (Tarjem 2022), institutional and disciplinary challenges persist, particularly norms, practices and biases regarding what constitutes knowledge and "good science" (Pyburn and van Eerdewijk 2021).

While scientification in the form of tool development may have enabled social scientists to gain some kind of foothold within agricultural development organizations, it has had important consequences for the nature of their contributions. Tools can reduce complexity and lower barriers for including gender in project design. By extracting data from the messiness of the social world, such information is "rendered technical," authorizing knowledge in accordance with dominant scientific paradigms (Li 2007). In presenting simplified data, tools make complex situations seem comprehensible. The information they produce is then used to identify and design solutions. Framed as technical outputs, they promise greater acceptance of social science perspectives.

However, while the production of simplified tools may seem to make aspects of social science research understood, in the process, deeper understanding of the complexities of gender systems is often lost: "In needing to provide an orderly route map for busy people, (tools) exclude content and complexity and become banal and mythic" (Standing 2004, p. 87). As Akter et al. (2017) highlight in their qualitative application of the WEAI tool in four Southeast Asian countries, although gender systems around the world are diverse and complex, purely quantitative tools often fail to capture cross-cultural variations in gender-specific needs and constraints, as well as the more nuanced and intangible aspects of gender relations. The simplification of social science perspectives through the use of generic gender analysis tools can lead to "glaring disconnects between contemporary understandings of gender and empowerment, and how they are commonly conceptualized and operationalized in programmatic agricultural development agendas" (Tavenner and Crane 2022, p. 855).

There is a danger that by focusing on tool development, researchers neglect broader research goals and objectives, as well as the socio-political agendas governing these. Almost 25 ago, Valerie Janesick (1998) raised concerns about "methodolatry," warning social scientists that cultivating and outlining their methods and procedures could result in researchers losing sight of the subject matter and thus generally undermining the potential of their research. In the case of gender analysis, while many organizations and projects have broad strategies and policies, and a plethora of gender analysis tools, they are often not clear on the focus of the research, required actions and expected outcomes (Njuki 2016).

If the overall research framing, aims and objectives are not clear, as Colverson et al. (2020) point out their study of the WEAI / WELI tools, researchers can end up measuring empowerment and collecting data without questioning the assumptions driving development initiatives, for example whether livestock intensification empowers or disempowers women. Such approaches risk missing the point on fundamental dimensions of what empowerment and gender relations mean in diverse cultural contexts, the foundations of gender disparities, and thus what gender-oriented development interventions intend to achieve (Tavenner and Crane 2022).

Despite the limitation of tools in achieving gender equity, multi- and bi-lateral donors, capitalist-philanthropists and other donors have intensified the demand for gender tools, perhaps contributing to the pressure within research institutions to produce them. In response, organizations have amassed a range of gender tools which have become part of their gender analysis toolboxes. Despite the many tools, some researchers and practitioners now describe gender tool development and use as performative (Bornstein 2006; Helleiner 2000). Most tools serve the purpose of project measurement and donor-recipient performance accountability, serving the demand for proof of social change and not messy reality (Batliwala and Pittman 2010, p. 7). They further argue that those who implement the tools are incentivized by sustaining or obtaining funding. These pressures, while appeasing funders in the short-term, mean that women's empowerment becomes a secondary concern to fundability and in the process the ability of development projects to contribute to meaningful change is constrained. These dynamics reflect broader structural power relationships among research institutions, development aid bureaucracies and philanthropic funders which are often far removed from democratic input, feminist political movements and from the people who are the stated beneficiaries of development efforts.

8.7 Politics of Instrumentalizing Social Science and Gender Analysis

Toolification of gender analysis has reduced the political project of gender and development to a technical fix. In the process, gender analysis has become ahistorical, de-politicized and decontextualized, leaving prevailing and unequal power relations intact. As such, the political dimensions of gender analysis have been rendered technical, narrowing the scope for transformative change (Cornwall et al. 2004, p. 4). As Yanow notes, the toolbox metaphor implies that "a researcher just needs a 'tool', any will do, whatever comes to hand," tools are "to be drawn on, at random, by anyone, in application to any random research question" (2012, pp. 38–39). This implies that anyone can do social science (or gender analysis). If you have the right tool you do not need special skills, knowledge, orientations, or training.

However, it is increasingly clear that when those implementing gender analysis tools do not have knowledge of the theory or disciplinary perspectives that they are grounded in, this contributes to the predominance of instrumental work that services the development industry in de-politicized ways (Mama 2007). By producing simplified tools as a response to the pressures they face, to prove their value and validity, and to make their work understandable and scalable, social scientists are simplifying and dumbing down the contributions they make. Power relations between research organizations and donor organizations—shaped by dominant political-economic agendas—shape the demands placed on social scientists in the AR4D sector. This reality indicates a need to understand the nested power relations which constitute the pressures on scientists to prove their scientific validity and economic value.

The metaphor of the toolbox suggests a departure from the reflexivity of the social sciences. Within social science disciplines, there is a recognition that research is ideologically driven, and that there is no value- or bias-free approach. Influenced by feminist scholarship, it is also widely acknowledged that research methods are social constructions, and social practices in their own right, that need to be reflexively interrogated (Tickamyer and Sexsmith 2019). Without this critical reflection, tools become thingified and become an end in themselves; they begin to exist in isolation, rather than being meaningfully integrated within research endeavors or effectively mobilized in pursuit of social change objectives.

The overreliance on standardized and simplified gender analysis tools within the agricultural research for development sector is arguably part of a McDonaldization trend within the social sciences more broadly (Ritzer 2011). As Nancarrow et al. observe, there seems to be a growing orientation towards standardized research practices, "just as McWorld creates a common world taste around logos, advertising slogans, stars, songs, brand names, jingles and trademarks ... the research world also seems to be moving towards a common world taste for an instantly recognizable and acceptable research method that can be deployed fast" (2005, p. 297). In the case of gender analysis, this can be seen in the "travelling circus of experts – gender technocrats touting a new kind of export product, whose brand name has shifted with the decades" (Mama 2004, p. 121), often promoting globalized notions of what gender equity looks like.

As feminist scholars have stressed over the years, ultimately, addressing gender inequality is a political undertaking, but often development organizations are uncomfortable with "radical" agendas, particularly the language and aims of feminism. Nicoline de Haan, Director of the CGIAR GENDER platform explained, "Generally, there is a hesitancy among our technical specialists to deal with social issues, which explains the fear of doing gender work especially in the context of the CGIAR where traditionally, the chief concern and goal was to increase (crop) productivity through scientific means" (Resurrección and Elmhirst 2021, p. 174).

This seemingly apolitical stance is reflected in the predominance of gender analysis tools that focus on data collection to inform narrow, technical agendas. As Merry explains, such "technocratic knowledge seems more reliable than political perspectives in generating solutions to problems, since it appears pragmatic and instrumental rather than ideological" (2016, p. 4). Of course, this denies the fact that all research and development is political, whether it intends to be or not. There is a consistent failure to deal with politics, complexity, context specificity and the dynamics of power relations within the development sector, and this is particularly acute within agricultural development organizations, partly because of the relative importance that different disciplinary perspectives are given (Cornwall et al. 2007, p. 12).

Focusing on tool development is convenient because it draws attention away from the fact that poor results in addressing gender inequality are often more to do with disparities in power and resources (March et al. 1999). While organizations have argued that a lack of appropriate tools prevent them from achieving their objectives, this is something of a toolbox fallacy, an assumption that you can only accomplish your goals if you have the right tools. While tools can be helpful, they do not guarantee results, particularly as most gender tools give no direct guidance in how to achieve desired development outcomes. As March et al. (1999, p. 14) highlight, before organizations can make informed choices on how they carry out gender analysis, they need to have clarity about the why and what their gender-specific objectives and strategies are trying to achieve. Ultimately, as Halsema (2003) points out, in terms of gender analysis it is the why that matters.

Tools alone will not fix agricultural research or change the way it is done. It could even be argued that tools can act as a distraction, shifting the emphasis away from critical social science and the transformational changes that such research might call for. Tools therefore exert their own agency in these arrangements, possibly hindering rather than facilitating structural change. Ultimately, tools cannot do the hard work of integrating local views into research and project design, nor can they put the "end users" of research and development processes into greater positions of power. As feminist social scientists have argued for decades, understanding local conceptualizations of gender equity and empowerment, and engaging participants in change processes, is a first step to designing more effective gender-transformative interventions.

8.8 Conclusion

Agricultural research for development requires a holistic understanding of the systemic structural and institutional barriers that reinforce gendered inequalities in agricultural contexts. As Mama (2007) emphasizes, attention to complexity, nuance, multiplicity and power relations is crucial, particularly in contexts where legacies of imperialism remain pervasive. In the case of agricultural development, this must extend to the organizations that carry out research and implement technologies. Addressing gender involves challenging existing organizational arrangements including how research questions and solutions are defined, allocations of authority and resources, as well as cultural and scientific norms. In these contexts, there is a critical need for investment in and support of adequate social science carried out by trained social scientists, rather than the tyrannical promotion of simplified, standardized tools that can seemingly be carried out by anyone in order to tick the boxes.

Rather than serving technical agendas determined by the biological sciences, the social sciences can contribute most effectively by playing a critical and reflexive role which entails questioning the objectives, methods and outcomes of research processes, and how they relate to broader socio-political agendas. This requires research institutions to increase social science capacity, particularly critically oriented social science disciplines, and not only the agricultural economists that currently prevail within the sector. Rather than being relegated to a service provision role, social scientists should have significant input in the framing and design of AR4D projects. A recent report by the Development Studies Association argues that, because an understanding of the social context (including research contexts) is critical to both the inputs and outcomes of development, "social science needs both to frame and ground interdisciplinary development research" (White 2019). This necessitates involvement in the definition of problems and generation of solutions, drawing on social science theories and methods.

Because research and development is political, which is particularly true when aiming to alter gender relations, there must be an acknowledgement that purely technical projects, particularly those conceptualized and implemented by outside "experts," will not produce meaningful change. Considering that questions of equity and empowerment have been extensively discussed and debated since the 1970s with little substantive change, despite the proliferation of gender analysis tools, it seems important to design alternative approaches and practices that deviate from the conventional paradigm (Hassanein 2000). An important step will be to work in collaboration with local and national organizations that already address gender equity on the ground. These partners are far better placed than external experts to facilitate transformation in the longer-term, beyond the typical two or three-year project cycle.

Kristjanson et al. (2017) make a similar argument, suggesting that joint learning is far more likely to result in determining the "why" and "so what" of research. This calls for more sustained engagement with stakeholders at all levels (from local upwards) and more effective and targeted communication with decision makers. Such transdisciplinary approaches can help to "challenge power structures and flatten hierarchies, situate knowledge production within contexts and relationships, and foster the co-production of knowledge as part of a social change process" (Elias et al. 2021 p 338). Ultimately, as feminist scholars advocate, meaningful attempts to facilitate gender empowerment and transformation need to be based on localized understandings that are relevant to people's lived realities, rather than concepts, tools and data that are externally constructed and applied by outsiders to meet normative scientific, donor and development agendas.

Appendix: Table of Tools

Tool	Citation	Year of origin	URL
Harvard Analytical Framework (Gender Roles Framework)	Overholt C, Anderson MB, Cloud K Austin JE (1985) Gender roles in development projects: a case book. Kumarian Press, West Hartford.	1985	https://pdf.usaid.gov/pdf_docs/Pnaas025.pdf
Capacities and Vulnerabilities Analysis (CVA) Framework	Anderson MB, Woodrow PJ (1989) Rising from the ashes: development strategies in times of disaster. Routledge, New York.	1989	https://www.adaptation-undp.org/sites/default/files/resources/6_capacities_and_vulnerabilities_assessment_framework_cva_framework.pdf
Longwe Framework	Longwe S (1991) Gender awareness: the missing element in the third world development project. In: Wallace T, March C (eds), Changing perceptions: writings on gender and development. Oxfam, Dublin.	1991	https://agris.fao.org/agris-search/search.do?recordID=GB9127251

Tool	Citation	Year of origin	URL
Moser Framework	Moser C (1993) Gender planning and development. Theory, practice and training. Routledge, Oxon.	1993	https://www.routledge.com/Gender-Planning-and-Development-Theory-Practice-and-Training/Moser/p/book/9780415056212
Gender Analysis Matrix (GAM)	Parker R (1993) Another point of view: a manual on gender analysis training for grassroots workers. UNIFEM, New York.	1993	https://www.ircwash.org/resources/another-point-view-manual-gender-analysis-training-grassroots-workers
Toolkit on Gender in Agriculture	Fong, MS, Bhushan A (1996) Toolkit on gender in agriculture. Gender toolkit series No. 1. World Bank, Washington DC.	1996	https://documents.worldbank.org/pt/publication/documents-reports/documentdetail/769711468739302708/toolkit-on-gender-in-agriculture
Project Cycle Management Technical Guide: Socio-Economic and Gender Analysis Programme	FAO (2001) Project cycle management technical guide. Socio-Economic and Gender Analysis Programme (SEAGA). Rome: Food and Agriculture Organization.	2001	https://www.fao.org/publications/card/fr/c/4de2b5da-e044-5d9e-91b2-70c8299dfe65/
Gender Checklist: Agriculture	Asian Development Bank (2000) Gender checklist: agriculture.	2006	https://www.adb.org/publications/gender-checklist-agriculture
Gender Action Learning System (GALS)	Mayoux L (2014) Gender Action Learning System (GALS): GALS Overview. Hivos, GALS@Scale project.	2007	http://www.galsatscale.net/_documents/GALSatScale0overviewCoffee.pdf
Gender Dimensions Framework	Rubin D, Manfre C, Nichols Barrett K (2009) Promoting gender equitable opportunities in agricultural value chains: a handbook. USAID, Washington DC.	2009	https://culturalpractice.com/resources/promoting-gender-equitable-opportunities-in-agricultural-value-chains-a-handbook/; http://hdl.handle.net/10919/69076
GAAP Gender and Assets Toolkit	Behrman J, Karelina Z, Peterman A, Roy S, Goh A, Kovarik C, Sproule K (eds) (2012) A toolkit on collecting gender & assets data in qualitative & quantitative program evaluations. International Food Policy Research Institute (IFPRI) and International Livestock Research Institute (ILRI), Nairobi.	2011	https://www.ifpri.org/publication/gaap-gender-and-assets-toolkit

Tool	Citation	Year of origin	URL
Gender Dimensions Framework: Gendered Perspectives for Conservation Agriculture	Harman Parks M, Christie, ME, Bagares I (2015) Gender and conservation agriculture: constraints and opportunities in the Philippines. GeoJournal 80:61–77.	2011	https://link.springer.com/article/10.1007/s10708-014-9523-4#citeas
Women's Empowerment Impact Measurement Initiative (WEIMI)	Berkowitz L, Gillingham S, Karim N Picard M (2014) Building capacity to measure long-term impact on women's empowerment: CARE's Women's Empowerment Impact Measurement Initiative. Oxfam, Routledge.	2010-12	https://policy-practice.oxfam.org/resources/building-capacity-to-measure-long-term-impact-on-womens-empowerment-cares-women-322259/
Engendered Chain Empowerment Matrix	KIT, Agri-ProFocus, IIRR (2012) Challenging chains to change: gender equity in agricultural value chain development. KIT Publishers, Royal Tropical Institute, Amsterdam.	2012	https://www.kit.nl/wp-content/uploads/2018/08/2008_chachacha.pdf
Methods for Mapping Gendered Farm Management Systems	Meinzen-Dick RS, van Koppen B, Behrman JA, Karelina Z, Akamandisa V, Hope L, Wielgosz B (2012) Putting gender on the map: methods for mapping gendered farm management systems in sub-Saharan Africa. IFPRI Discussion Paper 1153. International Food Policy Research Institute (IFPRI), Washington DC.	2012	www.ifpri.org/sites/default/files/publications/ifpridp01153.pdf
Women's Empowerment in AgricultureIndex (WEAI)	Alkire S, Meinzen-Dick RS, Peterman A, Quisumbing AR, Seymour G and Vaz A (2012) The Women's Empowerment in Agriculture Index. IFPRI Discussion Paper 1240. International Food Policy Research Institute (IFPRI), Washington DC.	2012	http://www.ifpri.org/book-9075/ourwork/program/weai-resource-center
Gender Action Learning System (GALS) (Agricultural Value Chains)	Reemer T, Makanza M (2014) Gender action learning system: practical guide for transforming gender and unequal power relations in value chains. Oxfam Novib.	2014	https://www.oxfamnovib.nl/Redactie/Downloads/English/publications/150115_Practical%20guide%20GALS%20summary%20Phase%201-2%20lr.pdf

Tool	Citation	Year of origin	URL
Climate Resilient Agriculture Tools	Jost C, Ferdous N, Spicer TD (2014) Gender and inclusion toolbox: Participatory Research in Climate Change and Agriculture. CGIAR Research Program on Climate Change, Agriculture and Food Security (CCAFS), CARE International and the World Agroforestry Centre (ICRAF), Copenhagen.	2014	https://cgspace.cgiar.org/ bitstream/handle/10568/45955/ CCAFS_Gender_Toolbox.pdf
Climate resilient agriculture module		2014	
Village Resource and Use Map		2014	
Goal Tree		2014	
Wealth & Vulnerability Ranking		2014	
Livelihood systems matrix		2014	
Perceptions of Women's Empowerment		2014	
Seasonal Calendar (Gender Roles)		2014	
Daily Activity Clocks		2014	
Changing Farming Practices		2014	
Venn Diagrams		2014	
Key Informant Interviews		2014	
Climate information module		2014	
Climate Information Ranking		2014	
Information networks game		2014	
Scientific forecasting		2014	
Mitigation Module		2014	
Changing farming practices timeline		2014	
Co-Benefit analysis		2014	

Tool	Citation	Year of origin	URL
How to do: Poverty targeting, gender equality and empowerment during project design: Gender, targeting and social inclusion	IFAD (2017) How to do: poverty targeting, gender equality and empowerment during project design: gender, targeting and social inclusion. International Fund for Agricultural Development (IFAD), Rome.	2017	https://www.ifad.org/ documents/38714170/41240 300/How+to+do+note+Poverty +targenting%2C+gender +equality+and+empowerment +during+project+design.pdf/0 171dde5-e157-4a6a-8e00- a2cafaa0e314
Gender in Irrigation Learning and Improvement Tool	Lefore N, Weight E, Rubin D (2017) Gender in irrigation learning and improvement tool. International Water Management Institute (IWMI), Colombo, Sri Lanka.	2017	http://www.iwmi.cgiar.org/ Publications/Other/training_ materials/gender_in_ irrigation_learning_and_ improvement_tool.pdf
G+ Tools for Gender Responsive Breeding	Ashby JA, Polar V (2021) User guide to the standard operating procedure for G+ tools (G+SOP). CGIAR Research Program on Roots, Tubers and Bananas, User Guide. 2021-3. International Potato Center: Lima.	2018	https://www.cgiar.org/ innovations/g-tools-for-gender- responsive-breeding/
GENNOVATE Ladder of Power and Freedom	Petesch P, Bullock R (2018) Ladder of power and freedom: qualitative data collection tool to understand local perceptions of agency and decision- making. GENNOVATE resources for scientists and research teams. CIMMYT, Mexico DF.	2018	https://gennovate.org/ wp-content/uploads/2018/10/ Ladder_of_Power_and_ Freedom_Gennovate_Tool.pdf
Project-level WEAI (Pro WEAI)	Malapit HJ, Quisumbing AR, Meinzen-Dick RS, Seymour G, Martinez EM, Heckert J, Rubin D, Vaz A Yount KM (2019) Development of the project- level Women's Empowerment in Agriculture Index (pro- WEAI). IFPRI Discussion Paper 1796. International Food Policy Research Institute (IFPRI), Washington, DC	2019	https://weai.ifpri.info/versions/ pro-weai/

Tool	Citation	Year of origin	URL
Abbreviated WEAI (A-WEAI)	Malapit HJ, Kovarik C, Sproule K, Meinzen-Dick RS, Quisumbing AR (2020) Instructional guide on the abbreviated Women's Empowerment in Agriculture Index (A-WEAI). International Food Policy Research Institute (IFPRI), Washington DC.	2020	http://ebrary.ifpri.org/cdm/ref/collection/p15738coll2/id/129719
Reducing the Gender Asset Gap through Agricultural Development	Quisumbing AR, Meinzen-Dick RS, Johnson NL, Njuki J, Julia B, Gilligan DO, Kovarik C, Peterman A, Roy S, Waithanji E, Rubin D, Manfre C (2014) Reducing the gender asset gap through agricultural development: a technical resource guide. Bangladesh Rural Advancement Committee (BRAC), CARE Cereal Systems Initiative for South Asia (CSISA), HarvestPlus, Helen Keller International, Kickstart International, Landesa, Land O'Lakes, International Food Policy Research Institute (IFPRI).	2020	http://gaap.ifpri.info/; http://www.ifpri.org/publication/reducing-gender-asset-gap-through-agricultural-development; http://genderassetgap.org/sites/default/files/ResearchBrief2.pdf.)

References

Acker J (1990) Hierarchies, jobs, bodies: a theory of gendered organizations. Gend Soc 4(2):139–158

Addison L, Schnurr MA, Gore C, Bawa S, Mujabi-Mujuzi S (2002) Women's empowerment in Africa: critical reflections on the abbreviated women's empowerment in agriculture index (A-WEAI). Afr Stud Rev 64(2):276–291

Akter S, Rutsaert P, Luis J, Htwe NM, San SS, Raharjo B, Pustika A (2017) Women's empowerment and gender equity in agriculture: a different perspective from Southeast Asia. Food Policy 69:270–279

Alkire S, Meinzen-Dick E, Peterman A, Quisumbing A, Seymour G, Vaz A (2013) The women's empowerment in agriculture index. World Dev 52:71–91

Amediume I (1987) Male daughters, female husbands: gender and sex in an African society. Zed Books, London

Ashby JA, Polar V (2021) User guide to the standard operating procedure for G+ tools (G+SOP). CGIAR Research Program on Roots, Tubers and Bananas, User Guide. 2021-3. International Potato Center, Lima

Barad K (2007) Meeting the universe halfway: quantum physics and the entanglement of matter and meaning. Durham University Press, Durham

Batliwala S, Pittman A (2010) Capturing change in women's realities: a critical overview of current monitoring and evaluation frameworks and approaches. Association for Women's Rights in Development (AWID), Toronto, Mexico City and Capetown

Bornstein L (2006) Systems of accountability, webs of deceit? Monitoring and evaluation in South African NGOs. Development 49:52–61

Boserup E (1970) Women's role in economic development. St Martin's Press, New York

Bourdieu P (1988) Homo academicus. Translated by Collier P. Stanford University Press, Stanford

Braidotti R (2021) Posthuman feminism. Polity Press, Cambridge

Brinkmann S, Jacobsen MH, Kristiansen S (2020) Historical overview of qualitative research in the social sciences. In: Leavy P (ed) The Oxford handbook of qualitative research, 2nd edn. Oxford University Press, Oxford, pp 17–42

Bryant L, Pini B (2006) Towards an understanding of gender and capital in constituting biotechnologies in agriculture. Sociol Rural 46(4):262–279

Caulkin S (2015) Feminism, interrupted? Gender and development in the era of "smart economics". Prog Dev Stud 15(4):295–307

Chambers R (2008) PRA, PLA and pluralism: practice and theory. In: The Sage handbook of action research. Participative inquiry and practice, 2nd edn. Sage, London, pp 297–318

Chant S (2016) Women, girls, and world poverty: empowerment, equality or essentialism? Int Dev Plan Rev 38(1):1–24

Colverson KE, Coble-Harris L, Galiè A, Moore EV, Munoz O, McKune SL, Singh N, Mo R (2020) Evolution of a gender tool: WEAI, WELI and livestock research. Glob Food Sec 26:1–6

Cornwall A (2016) Women's empowerment: what works? J Int Dev 28(3):342–359

Cornwall A (2018) Beyond "empowerment lite": women's empowerment, neoliberal development and global justice. Cadernos Padu 52:185–202

Cornwall A, Harrison E, Whitehead A (2004) Introduction: repositioning feminisms in gender and development. IDS Bull 35(4):1–10

Cornwall A, Harrison E, Whitehead A (2007) Gender myths and feminist fables: the struggle for interpretive power in gender and development. Dev Chang 38(1):1–20

Cummings S, Munthali N, Shapland P (2022) A systemic approach to the decolonisation of knowledge. In: Ludwig D, Boogaard B, Macnaghten P, Leewis C (eds) The politics of knowledge in inclusive development and innovation. Taylor & Francis, London, pp 65–79

Elias M, Cole S, Quisumbing A, Valencia A, Meinzen-Dick R, Twyman J (2021) Assessing women's empowerment in agricultural research. In: Pyburn R, van Eerdewijk A (eds) Advancing gender equality through agricultural and environmental research: past, present and future. IFPRI, Washington, DC, pp 329–364

Farhall K, Rickards L (2021) The "gender agenda" in agriculture for development and its (lack of) alignment with feminist scholarship. Front Sustain Food Syst 5:1–15

Foucault M, Rabinow P (1984) The Foucault reader. Pantheon Books, New York

Freire P (1970) Pedagogy of the oppressed. Seabury Press, New York

Gardner K, Lewis D (2015) Anthropology and development: challenges for the twenty-first century. Pluto Press, London

German L, Ramisch J, Verma R (eds) (2010) Beyond the biophysical: knowledge, culture and power in agriculture and natural resource management. Springer, Dordrecht

Halsema IV (2003) Feminist methodology and gender planning tools: divergences and meeting points. Gend Technol Dev 7(1):75–89

Haraway D (1991) Simians, cyborgs, and women: the reinvention of nature. Free Association Press, London

Harding S (1991) Whose science? Whose knowledge? Thinking from women's lives. Cornell University Press, Ithaca

Hassanein N (2000) Democratizing agricultural knowledge through sustainable farming networks. In: Kleinman DL (ed) Science, technology and democracy. State University of New York, Albany, pp 49–66

Helleiner G (2000) Towards balance in aid relationships: donor performance monitoring in low income developing countries. In: Dutt AK, Ross J (eds) Development economics and structuralist macroeconomics: essays in honor of Lance Taylor. Edward Elgar, Cheltenham, pp 336–351

Hickel J (2014) The "girl effect": liberalism, empowerment and the contradictions of development. Third World Q 35(8):1355–1373

Hodgson D (2001) Once intrepid warriors: gender, ethnicity, and the cultural politics of Maasai development. Indiana University Press, Bloomington

IFPRI (2012) A toolkit on collecting gender & assets data in qualitative and quantitative program evaluations. International Food Policy Institute, Washington, DC. https://ebrary.ifpri.org/utils/getfile/collection/p15738coll2/id/129042/filename/129253.pdf. Accessed 4 August 2022

Jackson C (2007) Resolving risk? Marriage and creative conjugality. Dev Chang 38(1):107–129

Janesick V (1998) The dance of qualitative research design: metaphor, methodolatry and meaning. In: Denzin NK, Lincoln YS (eds) Strategies of qualitative inquiry. Sage Publications, London, pp 35–55

Jasanoff S, Markle GE, Peterson JC, Pinch T (1995) Handbook of science and technology studies. Sage Publications, London

Jiggins J (1986) Gender-related impacts and the work of the international agricultural research centers. CGIAR Study Paper Number 17. The World Bank, Washington, DC

Kabeer N (1995) Targeting women or transforming institutions? Policy lessons from NGO anti-poverty efforts. Dev Pract 5(2):108–116

Kabeer N (1999) Resources, agency, achievements: reflections on the measurement of women's empowerment. Dev Chang 30:435–464

Kabeer N (2005) Gender equality and women's empowerment: a critical analysis of the third millennium development goal 1. Gend Dev 13(1):13–24

Kristjanson P, Bryan E, Bernier Q, Twyman J, Meinzen-Dick R, Kieran C, Ringler C, Jost C, Doss C (2017) Addressing gender in agricultural research for development in the face of a changing climate: where are we and where should we be going? Int J Agric Sustain 15(5):482–500

Latour B (1987) Science in action: how to follow scientists and engineers through society. Oxford University Press, Oxford

Leach M, Nisbett N, Cabral L, Harris J, Hossain N, Thompson J (2020) Food politics and development. World Dev 134:1–19

Li TM (2007) Practices of assemblage and community forest management. Econ Soc 36(2):263–293

Longwe SH (1991) Gender awareness: the missing element in the Third World development project. In: March C, Wallace T (eds) Changing perception: new writings on gender and development. Oxfam, Oxford, pp 149–157

Malapit H, Quisumbing A, Meinzen-Dick R, Seymour G, Martinez EM, Heckert J, Rubin D, Vaz A, Yount KM, GAAP2 Study Team (2019) Development of the project-level women's empowerment in agriculture index (pro-WEAI). World Dev 122:675–692

Mama A (2004) Demythologising gender in development: feminist studies in African contexts. IDS Bull 35(4):121–124

Mama A (2007) What does it mean to do feminist research in African contexts? Fem Rev 98(1):e4–e20

March C, Smyth I, Mukhopadhyay M (1999) A guide to gender-analysis frameworks. Oxfam, Oxford

Merry SE (2016) The seductions of quantification: measuring human right, gender violence, and sex trafficking. The University of Chicago Press, Chicago

Moser C (1989) Gender planning in the Third World: meeting practical and strategic gender needs. World Dev 17(11):1799–1825

Moser C (1993) Gender planning and development: theory, practice and training. Routledge, London

Nancarrow C, Vir J, Barker A (2005) Ritzer's McDonaldization and applied qualitative marketing research. Qual Mark Res Int J 8(3):296–311

Neef A, Neurbert D (2011) Stakeholder participation in agricultural research projects: a conceptual framework for reflection and decision-making. Agric Hum Values 28:179–194

Njuki J (2016) Practical notes: critical elements for integrating gender in agricultural research and development projects and programs. J Gender Agric Food Secur 1(3):104–108

Okali C (2012) Gender analysis: engaging with rural development and agricultural policy processes. (Working Paper 026). Future Agricultures Consortium. IDS, Brighton

Overholt C, Anderson MB, Cloud K, James AB (1985) A case book: gender roles in development projects. Kumarian Press, West Hartford

Oyewumi O (1997) The invention of women: making an African sense of Western gender discourse. University of Minnesota Press, Minneapolis

Polar V, Teeken B, Mwende J, Marimo P, Tufan HA, Ashby JA, Cole S, Mayanja S, Okello JJ, Kulakow P, Thiele G (2022) Building demand-led and gender-responsive breeding programs. In: Thiele G, Friedmann M, Campos H, Polar V, Bentley JW (eds) Root, tuber and banana food systems innovations: value creation for inclusive outcomes. Springer, Cham, pp 483–512

Pyburn R, van Eerdewijk A (2021) CGIAR research through an equality and empowerment lens. In: Pyburn R, van Eerdewijk A (eds) Advancing gender equality through agricultural and environmental research: past, present and future. IFPRI, Washington, DC, pp 1–75

Rathgeber EM (1990) WID, WAD, GAD: trends in research and practice. J Dev Areas 24(4):489–502

Rathgeber EM (2006) Who are the social researchers of the CGIAR system? In: Cernea MM, Kassam AH (eds) Researching the culture in agri-culture: social research for international development. CABI Publishing, Wallingford, pp 51–77

Razavi S, Miller C (1995) From WID to GAD: conceptual shifts in the women and development discourse. Occasional Paper no. 1. UN-RISD, Geneva

Resurrección BP, Elmhirst R (2021) Lifting the barriers of gender integration in livestock production. In: Resurrección BP, Elmhirst R (eds) Negotiating gender expertise in environment and development. Routledge, New York, pp 171–183

Rhoades RE (2006) Seeking half our brains: constraints and incentives in the social context of interdisciplinary research. In: Cernea MM, Kassam AH (eds) Researching the culture in agri-culture: social research for international development. CABI Publishing, Wallingford, pp 403–420

Ritzer G (2011) The McDonaldization of society. Sage Publications, London

Russo S (2012) Report for livestock climate change CRSP gender analyses of the LCC CRSP Portfolio. Livestock-Climate Change Collaborative Research Support Program, University of Florida

Selg P, Ventsel A (2020) The "relational turn" in the social sciences. In: Selg P, Ventsel A (eds) Introducing relational political analysis: political semiotics as a theory and method. Palgrave Macmillan, Cham, pp 15–40

Simons A, Vobs JP (2018) The concept of instrument constituencies: accounting for dynamics and practices of knowing governance. Polic Soc 37(1):14–35

Springer E (2020) Bureaucratic tools in (gendered) organisations: performance metrics and gender advisors in international development. Gend Soc 34(1):56–80

Standing H (2004) Gender, myth and fable: he perils of mainstreaming in sector bureaucracies. IDS Bull 35(4):82–88

Star SL (1990) Power, technology and the phenomenology of conventions: on being allergic to onions. Sociol Rev 38(1):vi–273

Tarjem IA (2022) Tools in the making: the co-construction of gender, crops and crop breeding in African agriculture. Gend Technol Dev 27:1–21. https://doi.org/10.1080/0971852 4.2022.2097621

Tarjem IA, Westengen OT, Wisborg P, Glaab K (2022) "Whose demand?" The co-construction of markets, demand and gender in development oriented crop breeding. Agric Hum Values 40:83–100. https://doi.org/10.1007/s10460-022-10337-y

Tavenner K, Crane T (2022) Hitting the target and missing the point? On the risks of measuring women's empowerment in agricultural development. Agric Hum Values 39:849–857

Tickamyer AR, Sexsmith K (2019) How to do gender research? Feminist perspectives on gender research in agriculture. In: Sachs CE, Jensen L, Castellanos P, Sexsmith K (eds) Routledge handbook of gender and agriculture. Routledge, New York, pp 57–71

Tsikata D (2001) Gender training in Ghana: politics, issues and tools. Woeli Publishing Services, Accra

van der Burg M (2020) Gender integration in international agricultural research for development. In: Sachs CE, Jensen L, Castellanos P, Sexsmith K (eds) Routledge handbook of gender and agriculture. Routledge, New York, pp 69–84

Verma R, Russell D, German L (2010) Anthro-apology? Negotiating space for interdisciplinary collaboration and in-depth anthropology in the CGIAR? In: German L, Ramisch J, Verma R (eds) Beyond the biophysical: knowledge, culture and power in agriculture and natural resource management. Springer, Dordrecht, pp 257–281

Vogel I (2012) Review of the use of "theory of change" in international development. Review Report for the UK Department of International Development. https://www.theoryofchange.org/pdf/DFID_ToC_Review_VogelV7.pdf

Wajcman J (2010) Feminist theories of technology. Camb J Econ 43(1):143–152

Warren H (2007) Using gender-analysis frameworks: theoretical and practical reflections. Gend Dev 15(2):187–198

West S, Schill C (2022) Negotiating the ethical-political dimensions of research methods: a key competency in mixed methods, inter- and transdisciplinary, and co-production research. Humanit Soc Sci Commun 9(2094):1–13

White S (2019) Towards more equitable interdisciplinary development research: five key messages. Development Studies Association

Wilson K (2015) Towards a radical re-appropriation: fender, development and neoliberal feminism. Dev Chang 46(4):803–832

Yanow D (2012) Organizational ethnography between toolbox and world-making. Organ Ethnogr 1(1):31–42

Chapter 9
An Intersectional Approach to Agricultural Research for Development (AR4D)

Katie Tavenner, Todd A. Crane, Renee Bullock, Alessandra Galiè, Hugo Campos ⓘ, and Gerald Katothya

Abstract Originating nearly 40 years ago in black feminist thought, the concept of intersectionality has become established as an analytical lens and social theory to account for and better understand multiple and compounding identities and how they influence discrimination and privilege. Within agricultural research for development (AR4D), intersectional approaches are relatively novel compared to traditional gender and social analyses, and to date there are limited tools and empirical studies in AR4D that have adopted such an approach. Without a strong conceptual and methodological foundation, future intersectional approaches in AR4D risk treating multiple identities as standalone "tick box" variables, and not as a holistic way of understanding and addressing these multiple sources of marginalization. To emphasize the potential value-addition of deeper engagement with intersectionality, this chapter outlines the state-of-the-field on intersectional analyses in AR4D and how they are situated within wider gender mainstreaming in international development. Using an empirical case study on index-based livestock insurance (IBLI) in Northern Kenya, the chapter demonstrates an intersectional analysis in AR4D, based on a new conceptual framework and method (Tavenner et al. Gend Technol Dev 26(3):385–403, 2022). This chapter explores how AR4D can deepen its understanding of intersectionality and the potential integration of this concept in a meaningful way that supports addressing multiple layers of inequalities and marginalization in agricultural research methods and practice.

K. Tavenner (✉)
Independent Consultant, London, UK

T. A. Crane · R. Bullock · A. Galiè
International Livestock Research Institute, Nairobi, Kenya
e-mail: t.crane@cgiar.org; r.bullock@cgiar.org; a.galie@cgiar.org

H. Campos
International Potato Center, Lima, Peru
e-mail: h.campos@cgiar.org

G. Katothya
Independent Consultant, Nairobi, Kenya

© The Author(s) 2025   167
J. Njuki et al. (eds.), *Gender, Power and Politics in Agriculture*,
https://doi.org/10.1007/978-3-031-60986-2_9

9.1   Introduction

Since its origin almost 40 years ago in black feminist thought (Crenshaw 1989), the concept of intersectionality has emerged as an analytical lens and social theory to account for and better understand multiple and compounding identities and how they influence marginalization and privilege. While the definition and conceptualization of intersectionality as an analytical lens have been debated by feminist scholars from the start (Cho et al. 2013; Choo and Ferree 2010; Hancock 2007; McCall 2005; Collins 2002), the focus of this chapter is on the operationalization and application of an intersectional approach in the context of agricultural research for development (AR4D).

An intersectional perspective is relevant to AR4D because it recognizes the heterogenous nature of people facing multiple and intersecting forms of marginalization (Lotfata and Munenzon 2022; Ryder 2017). In the case of women, it recognizes and addresses the role of diverse intersections and of power hierarchies, often a significant gap in development efforts. The use of general categories such as "women and girls" can miss the point about the myriad and intersecting ways of discrimination and inequality, leaving power issues sidelined (Lokot and Avakyan 2020). An intersectional perspective in A4RD highlights nuances well beyond the apparent ones. For instance, using an intersectional approach to analyze gendered pathways mediating access to water, and the consequences for nutrition and gender equality in two districts in India, Mitra and Rao (2019) were able to pinpoint how caste, ethnicity, and class interact to shape dependence on particular water sources. This provides useful insights about the linkages between the availability of water, its uses, food and nutrition security, and gender roles and responsibilities.

Within the AR4D field, intersectional approaches are relatively novel compared to traditional gender and social analyses and to date there are limited tools and empirical studies in AR4D that have adopted such an approach (notable, excellent exceptions include Badstue et al. 2020; Clement et al. 2019; Leder et al. 2017). Without a strong conceptual and methodological underpinning, future intersectional studies in AR4D risk treating multiple identities as standalone "tick box" variables, and not as a holistic way of understanding and addressing these multiple and intersecting sources of marginalization (Hunting and Hankivsky 2020). Yet, there are few practical analytical approaches developed to capture the compounding effect of multiple social characteristics that shape the lived experiences of individuals. Approaches are also needed to understand the patterns that are common among these unique experiences of marginalization to identify the main determinants of power hierarchies.

To emphasize the potential value-addition of deeper engagement with intersectionality, this chapter outlines the state-of-the-field on intersectional analyses in AR4D and how they are situated within wider gender mainstreaming in international development. These gender-mainstreaming domains in AR4D include gender-transformative approaches (GTA), measurements of women's empowerment, and decolonizing development approaches. The chapter also discusses the trade-offs in adopting an intersectional approach against current AR4D systematic priorities of scaling and mass impact.

Using an empirical case study on index-based livestock insurance (IBLI) in Northern Kenya, the chapter demonstrates the application of intersectional analysis in AR4D based on a new conceptual framework and methodological approach (Tavenner et al. 2022). The chapter explores how AR4D can deepen its understanding of intersectionality and the potential application of this concept in a meaningful way that addresses multiple layers of inequalities and marginalization in agricultural research methods and practice.

We acknowledge that in operationalizing "intersectionality" at field-level there is a risk of instrumentalizing, depoliticizing, and diluting the concept into a narrow development practitioner tool—similar to feminist critiques on "women's empowerment" in development programming (Tavenner and Crane 2022). At the same time, we assert that in attempting to operationalize the concept, intersectional analysis can illuminate and deepen understanding of multiple and intersecting layers of discrimination and marginalization that mediate people's participation in, and benefit from agriculture—insights that are valuable for development practitioners.

We present this chapter with the recognition that we as authors need to reflect on our own intersectional positionality. Four of the six authors are Euro-American researchers that have benefited from systemic inequalities in higher education and international development space via lingering effects of colonialism and racism. We welcome the opportunity to actively reflect on this positionality, and act toward decolonizing AR4D, to which we hope this chapter can contribute. In applying this specific conceptual framework and method, we also acknowledge that:

> Intersectionality ... does not provide written-in-stone guidelines for doing feminist inquiry ... instead it encourages each feminist scholar to engage critically with her own assumptions in the interests of reflexive, critical, and accountable feminist inquiry (Davis 2008, p. 79).

9.2 Framework and Method for an Intersectional Approach to Gender and AR4D

9.2.1 Conceptual Framework

Operationalization has historically been a challenge in translating the complexities of feminist concepts and theories into actionable development planning and progress. For example, development practitioners have posited that the transformative potential of targeting "empowerment" as a broader development goal has, at the intervention level, largely been watered down to fit narrow neoliberal development aims (Calkin 2015; Chant and Sweetman 2012). This makes "empowerment" a weakly implemented development concept (Tavenner and Crane 2022), or at its worst, a virtue-signaling buzzword (Cornwall and Brock 2005). To help intersectionality avoid a similar fate, moving it forward as a meaningful concept within the feminist agenda of AR4D institutions, requires that gender researchers and practitioners advocate for its value-addition as an applied concept (i.e., providing richness, nuance, and context to discuss how varying social identities are tied to broader structural power relations within agricultural systems). Methodological guidelines are also needed on how intersectionality can be pragmatically operationalized, implemented, and monitored in AR4D interventions. Developing research design guidelines based on concrete concepts and tools and piloting them is essential to moving substantive intersectional research practice forward. Table 9.1 provides a brief list of the key concepts that underpin the incorporation of intersectionality in gender and agriculture.

Although sex is the most frequently used variable to assess inequalities, particularly among women and men, and boys and girls, there are additional structural and systemic factors shaping their opportunities and agency in political, social, and economic aspects (Ng'endo and Connor 2022).

Social Embeddedness By outlining the rationale for how intersectionality can inform AR4D, we draw on the sociological theory of social embeddedness (Granovetter 1985), which theorizes that agricultural activities, technologies, and institutions are positioned *within* (rather than outside of) socio-cultural systems. The engagement of individual actors within AR4D interventions are mediated by social relationships (at household and community levels), norms (including, but not limited to hegemonic gender norms), and meanings ascribed to different activities, value chains, and agricultural products are embedded in historically and geographically specific socio-cultural milieus that structure people's engagement with them (Tavenner et al. 2022, p. 389).

Table 9.1 Key concepts and definitions

Key concept	Brief definition
Gender	The socially-constructed system of classification that ascribes qualities of masculinity and femininity to people, often based on their biological sex. Gender characteristics can change over time and are different between cultures (Vinyeta et al. 2015).
Intersectionality in gender and agriculture	The ways that gender inequalities intersect with other forms of inequality/ axes of social differentiation (such as age, assets base, marital status, race/ ethnicity, and caste/class) within specific historical and cultural contingent contexts to influence people's ability to benefit from agricultural development (Ravera et al. 2016). Such multiple intersects in turn define power relationships.
Gender-transformative approaches (GTA)	Approaches that seek to tackle the structured root causes of entrenched gender inequalities at multiple scales, including gender norms and roles, rather than merely responding to the symptoms of gender inequality that such structures produce (Farhall and Rickards 2021).
AR4D	Agricultural research for development—refers to agriculture-based international development efforts (Farhall and Rickards 2021) which typically target the needs of smallholder farmers of the Global South.

Therefore, connecting intersectionality with a theory of social embeddedness in agriculture frames intersecting identities and axes of social difference (e.g., gender, age, assets base, marital status, race/ethnicity, and caste/class) as layered power dynamics that structure an individual's participation, opportunities, and constraints in agricultural production (Tavenner et al. 2022). While these power dynamics are patterned by history and culture (Ravera et al. 2016) and are experienced differently by individuals, they do demonstrate broader patterns of gendered farmer engagement—for example, by determining the social acceptability for women and men to engage in farm activities that do not align with their gender. Ultimately, connecting intersectionality with a theory of social embeddedness shows how power relations along different axes of identity confer advantage or discrimination among agricultural actors.

Such understanding can help make progress towards both agricultural development (by showing for example, what dynamics need tackling for agricultural innovations and technologies to be adopted by all farmers) and gender equality (by ensuring, for example, that no one is disadvantaged by the introduction of such innovations, but rather, such innovations are leveraged to reduce disadvantage). Adopting such an approach supports inclusive gender-transformative approaches (GTA) which analyze and address the root causes of gender inequalities, as opposed to merely addressing the symptoms of gender inequality (Farhall and Rickards 2021). Indeed, intersectionality is an inherent concept in GTA, and crucial in identifying which interactive social layers are the most relevant to mediating opportunities and constraints in agriculture. Intersectional analysis can provide crucial insights into how AR4D can be more socially inclusive, and therefore more effective, in working towards GTA goals. Intersectional analysis can help to operationalize gender research outputs, to go beyond research, and actually address gender gaps.

We adopt the approach outlined by Tavenner and Crane (2019), which draws on Choo and Ferree's (2010) seminal work on understanding intersectionality in practice. Choo and Feree suggest that intersectionality can be analyzed using three styles: group-centered, process-centered, and system-centered. The group-centered approach places marginalized groups and their perspectives at the center of the research. A group-centered approach illuminates hidden social groups and their barriers to participation, benefit, and empowerment in AR4D.[1] The process-centered approach views power as relational, and can highlight how livelihood practices and development processes intersect with gendered social groups and power relations to influence AR4D outcomes. A system-centered style views intersectionality as structuring entire social systems that are historically situated and complex (Choo and Ferree 2010). A system-centered lens can explore how agency intersects with institutional and structural constraints and systems to shape farmer self-perceptions (Tavenner and Crane 2019). Identifying power dynamics embedded in social structures and institutions is also at the center of GTA.

---

[1] We use the term "social group" to refer to people facing similar multiple and intersecting forms of marginalization.

9.2.2   State-of-the-Field on Intersectional Analyses in AR4D

This section provides a brief state-of-the-field on intersectional analyses in AR4D and how they are situated within wider gender mainstreaming in international development. These include gender-transformative approaches (GTA), measurements of women's empowerment, and decolonizing development approaches.

Intersectionality in AR4D Has Been Framed Through the Lens of Gender Mainstreaming A recent scoping study by the Adaptation Fund (2022) provides an in-depth assessment of the historical trajectory of the concept of intersectionality in adaptation-relevant interventions, including those in the agriculture sector. The study demonstrates how the utility of intersectionality has been explained mainly through gender-mainstreaming agendas. Framing intersectionality in AR4D via gender mainstreaming reflects some transition from ideas arising in radical, subaltern feminisms towards praxis and development policy implementation (Patil 2013).

The Most Common Intersectional Relationship Examined in AR4D Is Gender and Age While there is growing diversity in the agricultural literature on different types of intersectional relationships, gender and age are the most prevalent (see McKune et al. 2021 for a full list of studies). This is because "age" is a proxy for the broader development narrative around "youth." As noted by Tavenner and Crane (2019), "women and youth" are commonly bundled in development discourse to denote commitments to gender equality and social inclusivity. For example, the intersections of gender and age have been investigated in terms of creating opportunities with dairy intensification in Kenya (Bullock and Crane 2021) and how gender and age interact with wider community changes to inform opportunities for and constraints on the production of value-addition agricultural commodities (Tavenner and Crane 2019).

The Growing Body of Literature on Identifying Potential Intersectional Impacts in AR4D Interventions Empirical studies in AR4D have explored how social groups use different climate-change adaptations in farming (McKune et al. 2021; Thompson-Hall et al. 2016), including by women farmers in Cameroon (Ngum and Bastiaensen 2021) and by women fish traders in Cambodia (Kusakabe and Sereyvath 2015). Research in AR4D has used gender to analyze environmental resources in Nepal (Nightingale 2011) and how indigeneity, gender, and class have advanced food well-being through intersectionality (Ashik et al. 2022).

Intersectional environmental studies also have implications for AR4D (Kaijser and Kronsell 2014), for example, studying how the interactions between gender, caste, and class, influence decision-making on risks from natural resource extraction (Kojola 2019). In fisheries, commonly neglected as an agri-food system, the intersections of gender, marital status, nationality, and migration status affected seafood traders in Palau (Ferguson 2021). A recent global review of ocean policy interventions identified wealth, marriage, family roles, and social networks as having possible intersectional impacts (Axelrod et al. 2022).

Other recent additions to the practical applications in AR4D include the intersections of gender with marital status in accessing climate adaptation information in Tanzania (Van Aelst and Holvoet 2016), the intersections of gender with caste and women's empowerment in dairy cooperatives in India (Ravichandran et al. 2021; Farnworth et al. 2023), and intersections of gender with race and class in influencing urban agricultural practices in the United States (Whitley 2020).

New Toolkits on Incorporating Intersectional Analysis in Projects and Programs Related to Climate Change and Agriculture These include: a toolkit targeting forestry, but with applications to agroforestry food systems (Colfer et al. 2018). A toolkit for integrating intersectionality in sustainable rural development programs (FAO 2023). And a toolkit for understanding intersectional perspectives on local climate priorities, particularly through the lens of gender and age in Tanzania (Greene et al. 2020).

Intersectionality Is an Essential Component of GTA and Aligns with Decolonizing Development Research Efforts but Complicates Current Metrics for Measuring Women's Empowerment Although not always presented as such, intersectionality is a fundamental concept underpinning GTA (Oosthuizen 2023). Without attention to the multi-faceted dimensions of inequalities that pattern women's and girl's lives across diverse backgrounds, GTA's goal of addressing and dismantling the gender norms and roles underlying inequalities cannot be addressed. Intersectionality also aligns with efforts to decolonize development research by way of recognizing institutional and structural racism in agricultural research organizations (Gewin 2022). However, adopting an intersectional approach complicates metrics for measuring women's empowerment that generally rely on binary sex-disaggregated data and do not consider how gender power dynamics are understood and constructed locally (Tavenner and Crane 2022).

The next section details a 5 step method on how to operationalize intersectionality in AR4D, to go beyond binary gendered approaches in designing interventions. The flexible steps provide a suggested order, but they can be adapted to different contexts.

9.2.3 Method

Any research design method should guide the research process. Since the intersectional approach in AR4D is new, and it approaches and best practices are still being developed, we draw on the hypothetical method outlined in Tavenner et al. (2022), illustrated in 5 steps in Fig. 9.1. The steps form a holistic approach that includes how to formulate key analytical questions, identify relevant intersectional attributes, select appropriate methods, sequence research activities, engage in analysis informed by intersectionality, and capture intersectional analyses in ways that can inform AR4D interventions and programmatic thinking.

The first step, "Formulating the key analytical questions to identify the main axes of social differentiation and the way they are embedded in the agricultural system", is the foundation for research and sets the stage for the scope and purpose of the AR4D. This step builds the framework for analysis of intersectionality in local contexts, identifying which axes of social differentiation exist and how they are socially embedded in a specific agricultural system. This requires identifying the power structures, including informal ones (norms/beliefs/practices) and formal ones (rules/regulations) of different axes of social identity (Tavenner et al. 2022). Researchers should incorporate literature review, stakeholder consultations with partners and community members, and qualitative situational analysis (Seager 2021) to inform the process.

The method provides practical guidelines for decentering development practitioners' goals to focus more on how AR4D can help to achieve locally-defined goals. Step 1 requires researchers to step back and reflect on their own privilege, biases, positionality, and agendas and listen carefully to the people who may participate in the intervention to align the AR4D with the needs, and goals of the people involved.

Step 2, "Bounding" identifies which intersectional axes will be selected and explored in the analysis. Following on the participatory co-creation of appropriate analytical questions, the rationale for the selection of social axes should be grounded in the AR4D's goals as co-defined by local stakeholders, potential participants, and development practitioners. While goals cane be defined in several ways (e.g., community conversations, stakeholder meetings), it is of essence to hold many

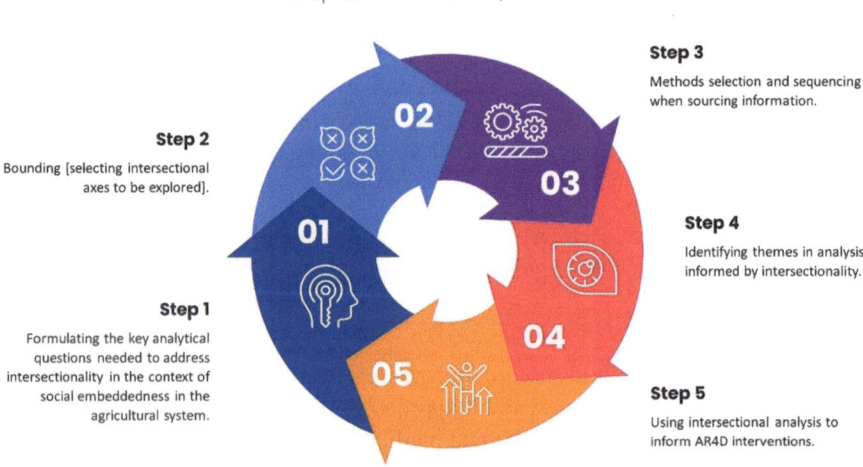

Step-by-step research design for applying intersectionality in AR4D

Adapted from Tavenner, et al. 2022

Step 3
Methods selection and sequencing when sourcing information.

Step 2
Bounding [selecting intersectional axes to be explored].

Step 4
Identifying themes in analysis informed by intersectionality.

Step 1
Formulating the key analytical questions needed to address intersectionality in the context of social embeddedness in the agricultural system.

Step 5
Using intersectional analysis to inform AR4D interventions.

Fig. 9.1 Step-by-step research design for applying intersectionality in AR4D. (Source: Tavenner et al. 2022)

discussions around community power dynamics that reinforce exclusions. These discussions are critical in answering the key bounding questions: What are the key social axes in each agricultural system? How are these axes linked? Are some social axes more relevant than others? Involving the most marginalized individuals in such discussions is essential to avoid reproducing structural disadvantage in the very definition of locally relevant goals, if they are formulated by the more privileged members of the community.

Step 3, "Methods selection and sequencing when sourcing information", is selection of appropriate qualitative and quantitative methods and deciding when to use them during the AR4D cycle. These decisions should consider how gender and social relations evolve in agricultural systems. The methods should be iterative, ethical, and provide plans for data collection, monitoring, and evaluation in ways that are not exploitative or antithetical to the co-creation of intersectional analytics (Tavenner et al. 2022).

Step 4, "Identifying themes in analysis informed by intersectionality" identifies specific themes based on the research questions and methods used. The themes depend on the research questions and methods, but the analysis should cover the three styles of intersectionality described by Choo and Ferree (2010): group-centered, process-centered, and system-centered. These styles can be captured by clustering results around three analytic areas: (1) the barriers to participation, benefit, and empowerment for specific (and potentially hidden) social groups; (2) how livelihood practices and development processes intersect with social groups and power relations to influence outcomes; (3) and how agency intersects with institutional and structural constraints in systems to shape self-perceptions (Tavenner and Crane 2019).

Step 5, 'Using intersectional analysis to inform AR4D interventions' describes how the intersectional research findings can be applied to improve future AR4D interventions, including how intersectionality can inform GTA in broader AR4D agendas. Using such a lesson learnt approach (Sverdlik 2021) can help unpack researcher reflections on the process of using the methodology, and on collating recommendations for future intersectional research.

9.3 Empirical Case Study: An Intersectional Approach to Understand Decision-Making About Livestock Insurance Uptake in Northern Kenya

This section uses the method for an applied intersectional approach in AR4D, outlined in Tavenner et al. (2022), via a case study of livestock insurance in Northern Kenya. Case study data was collected in two rounds in July and November 2022. The method was applied retroactively between December–February 2023 due to inadequate time and resources to apply the approach before publication.

Index-Based Livestock Insurance Over the last decade, the International Livestock Research Institute (ILRI) and its partners and donors, have introduced a program dedicated to index-based livestock insurance (IBLI—see https://ibli.ilri. org). IBLI was first launched in the arid and semi-arid lands of Kenya in 2010, and in Ethiopia in 2012. Over time, the IBLI program has evolved and been scaled out as Drought Risk Financing Solutions (DRFS). Initially a microinsurance product, DRFS are now considered to be an important social protection program that can improve households' resilience to climate shocks and reduce risk among pastoralists in the Horn of Africa (HoA), also known as the Somali Peninsula.

Microlevel retail IBLI schemes have been implemented in Northern Kenya and Southern Ethiopia with private insurance companies involved in marketing, promoting, and underwriting the scheme on a voluntary basis with individual pastoralists to insure cattle, camels, sheep, and goats. The schemes have been active in Kenya since 2010 and in the Borena region of Ethiopia since 2012. The uptake of the product in both cases has been slow but constant and now about 7000 policies are sold per year. Between Kenya and Ethiopia, the cumulative value of IBLI purchases was over 40,000 Tropical Livestock Units (TLUs) valued at almost $6 million.

Macrolevel social livelihood protection insurance schemes are currently operational in Kenya and in Eastern Ethiopia, while a pilot program has been recently launched in Zambia (Machado and Good 2022). The macrolevel initiative was initiated by the Government of Kenya (GoK) in 2015 under the Kenya Livestock Insurance Program (KLIP), supported by a public-private partnership (PPP). KLIP premiums are fully funded by GoK and currently the program covers about 18,000 (each for five TLUs) of the most vulnerable pastoral households in eight arid and semi-arid counties of Kenya, representing over 80,000 beneficiaries. KLIP is intended to scale to 100,000 households across 16 counties by 2021. The 2016–17 drought was among the worst in Kenya in the past 20 years, and KLIP paid out $7 million to pastoralists. In Ethiopia, the World Food Program has been implementing a similar, fully subsidized scheme since 2018 under the Satellite Index Insurance for Pastoralists in Ethiopia (SIIPE) program, currently covering 15,000 households. However, with the evolving pastoral systems and threats of climate change, there have also been demands for DRFS to rise to several other challenges in the drylands of Africa, including community-level conflict, access to veterinary services, and climate-resilient fodder for livestock. These challenges are situated in the local context of gender inequality in access to resources, information, decision-making, and livestock assets.

Rationale for an Intersectional Approach This case study uses gender as the first axis of social difference in intersectional analysis. Gender is a useful entry point to understand how different axes of social differentiation and inequalities can create differential vulnerability in ways that mediate the decision-making, uptake and benefits of livestock insurance. For example, where information channels are male-dominated, women cannot decide to buy agricultural insurance, because they have no access to information about it. Women in sub-Saharan Africa (SSA) often have lower education and financial literacy, making it difficult for them to understand the agricultural insurance designs and compensation procedures (Timu and Kramer 2021). In SSA, women have constrained personal independence, limited resources and little decision-making power even on matters affecting their communities and their own lives (Fisher and Carr 2015). Thus, even if the women want to take up agricultural insurance, they cannot independently make that decision. Instead, they must consult their spouses or other leading men in the community (Timu and Kramer 2021).

In pastoralist systems, animal ownership is gendered and women's rights to livestock are often less secure than men's (Flintan 2021). While drought affects all members of pastoral households, women livestock keepers are often disproportionately affected because of their less secure claims to assets. Women often take out lower rates of agricultural and livestock insurance than men. For example, IBLI quantitative panel data show greater dis-adoption among women policyholders than men in Marsabit county, Kenya. Women policyholders, on average, pay lower premiums and insure fewer cattle than men, while both genders insure the same number of sheep and goats. The panel data suggest that there are other socioeconomic and locational influences on uptake of IBLI, beyond gender, so an intersectional lens is important for understanding intrahousehold level decision-making on IBLI.

Site background. Marsabit county is a vast semi-arid region in the extreme north of Kenya, bordering Ethiopia. The last national census estimates that 459,785 people in 77,495 households (KNBS 2019) are spread across four sub-counties: Laisamis, Marsabit Central/Saku, North Horr, and Moyale. Marsabit county is vulnerable to climate change and recurrent droughts. Conflicts between Borana and Gabra ethnic groups over pasture have been frequent since 2016,17. The majority ethnic groups in the study area are the Samburu (Laisamis sub-county), Borana (Marsabit Central/Saku sub-county), and the Gabra (North Horr sub-county).

The different livestock and gender norms of these ethic groups create different opportunities and constraints for women to own, manage, and benefit from livestock. Climate risk influences the species owned in different study sites. Men tend to own larger animals, such as cattle and camels, while women own goats and sheep. Poultry keeping is an increasing trend in all sites, managed by women. Livestock are central in livelihoods in the study sites and herders often migrate between base camps, or sedentary homes, and satellite camps, where pasture is available. Men and boys herd livestock while women remain in the homesteads, or the basecamps. Insecurity is exacerbated by cattle rustling and raids along migration routes.

9.3.1 Step 1: Formulating the Key Analytical Questions Needed to Address Intersectionality in the Context of Social Embeddedness in the Agricultural System

Our initial research question was: How do socioeconomic characteristics, including gender and ethnicity, influence uptake of insurance?

After discussions with field staff about social factors and stakeholders, we refined the question for the intersectional analysis to ask: How do gender, ethnicity, and household structure intersect and influence livestock insurance decision-making, specifically who participates in decision making about whether to take up insurance, who makes the final decision about whether to purchase insurance, and who participates in decision-making about pay-outs?

We systematically collected information about social relationships in different household structures (monogamous and polygynous) across the locations. We also collected information about major climate and crisis events, and norms of livestock ownership and decision-making.

9.3.2   Step 2: Bounding [Selecting Intersectional Axes to Be Explored]

The selection of intersectional factors (gender, ethnicity, and household structure) were informed by four activities: literature review, project panel data on dis-adoption, consultations with ILRI IBLI field staff, and key informant interviews. Ideally, consultations with those most marginalized are crucial in setting bounding priorities. However, a limitation in retrofitting an intersectional approach is that the team was limited to working with existing data.

The literature revealed a limited understanding of how gender interacts with livestock insurance uptake and benefit distribution, with implications for managing and reducing climate risk and shocks. Livestock insurance is primarily offered in pastoralist communities, where multiple compounding and complex uncertainties interact with sociocultural factors, like ethnicity and marriage practices, in significant, yet undocumented ways.

IBLI quantitative panel data revealed important trends in dis-adoption over time, but it was unclear how gender was related to uptake and benefit distribution. Discussions with IBLI staff and key informants provided richer insights on the roles of ethnicity and marital status. Field staff explained the prevalence of polygyny and the importance of ethnic conflict in the study sites.

In our analysis the household structure emerged as more important than age in the uptake of IBLI. Few youth participated in IBLI, because many young men had migration from the area. However, it may have been worthwhile to interview mothers of young men inheriting animals or requiring animals for marriage as social drivers. Investigating whether wives with more sons have more decision-making power in polygynous households would have been another point for consideration.

The quantitative panel data show that policyholders are often men and that women generally insure fewer and different species compared to men. The qualitative data explored these trends to identify the ways in which gender, ethnicity and marital status/household type (monogamous and polygynous marriage) affect product uptake to guide future efforts to support socially inclusive scaling of IBLI. The team investigated gender norms and practices related to livestock owner-ship and women's and men's decision-making to take up insurance and benefit dis-tribution. These activities and theorizing that gender is influenced by, and intersects with, factors such as ethnicity and household arrangements motivated the selection of the most relevant intersectional, and deliberately researched axes.

9.3.3   Step 3: Sequencing Methods When Sourcing Information

This step is presented in three parts: planning, data collection and analysis.

Part 1: Planning Site selection was informed by project- and county-specific characteristics. Marsabit county is relevant for the project and representative of arid and semi-arid lands (ASAL) in Africa where livestock-based livelihoods are com-mon, and drought is the primary climate hazard. IBLI has operated in Marsabit county (2010-current) and scaling is planned. Complementary quantitative panel data (2010–2020) also exists.

Sub-counties were selected to maximize social, ecological, market and livestock diversity (Table 9.2). Since ethnic conflicts are a sensitive topic in Marsabit we intentionally selected sites to sample ethnic groups, including Samburu, Borana and Gabra, that have different cultural norms and practices. The variation in agro-ecological zones (AEZ) shows how access to natural resources, e.g., forests, helps cope with climatic shocks.

Participants were selected by purposive sampling of individuals who had prior experience subscribing to IBLI to better understand perspectives and experiences with the product.

All data was collected in local language and facilitated through sex- and ethnically-disaggregated groups. Field teams participated in a training program that included translation of survey instruments from English to Swahili and into the local languages of Borana, Gabra and Samburu. The instruments were revised and reviewed through practice sessions, which allowed testing the draft questionnaires for a shared understanding of the questions in the three local languages.

Part 2: Data Collection Data was collected through seven key informant inter-views (KIIs)and 32 focus group discussions (FGDs) in three sub counties in Marsabit county between July and November 2022 (Table 9.3). The 32 sex disag-gregated FGDs (50% women) allowed the researchers to understand key historical events, gender roles and practices related to livestock, uptake and outcomes of IBLI, and changes during crisis events, such as conflict or drought. Each FGD was com-posed of seven to 12 members; a total 321 respondents (48% women) participated. Each FGD took about two and half hours, with simultaneous translation of Kiswahili instruments into Borana or Samburu languages. Seven KIIs were held with

Table 9.2 Site sampling selection criteria

Sites	Sub-county	Population	Main ethnic group	AEZ
Laisamis	Laisamis	74,131 (3.65/km²)	Samburu	Lowland/dry
Marsabit Mountains	Marsabit Central/ Saku	52,521 (25.6/km²)	Borana	Highlands/mixed wet & dry
Bubisa	North Horr	84,935 (2.16/ km²)	Gabra	Lowland/dry

Source of population data: 2020 projection

Table 9.3 Summary of qualitative study methods

Method	Women	Men	Total
Key informant interviews	0	7	7
FGDs-Round 1	6 (59)	6 (61)	12 (120)
FGDs-Round 2	10 (96)	10 (105)	(20)201
FGDs total	**16 (155)**	**16 (166)**	**32 (321)**
Total # participants	**155**	**173**	**328**

livestock officers, local IBLI agents, and a former manager with an insurance company offering IBLI.

Part 3: Data Analysis Qualitative data was analyzed using iterative deductive and inductive approaches. All transcripts were transcribed in English and imported into NVivo. A codebook was developed to guide the first round of coding by the researcher, which was then cross checked and discussed to ensure intercoder reliability. Next, themes related to gender and decision-making were identified. Qualitative comparative analysis (QCA) was performed both within and across study locations to identify trends and differences between sites, or between the ethnic groups (Mello 2021).

9.3.4 Step 4. Identifying Themes in Analysis Informed by Intersectionality

The themes identified by the intersectional analysis are presented below in relation to the adapted research question: How do gender, ethnicity, and household structure intersect and influence livestock insurance decision-making, specifically:

Who is involved?
Who decides to purchase?

9.3.4.1   Theme 1: Intersectional Analysis Helps Identify Hidden Social Groups and Barriers to Livestock Insurance Decision-Making

Key Finding In polygynous households, relationships between a husband and multiple wives and co-wife relationships (wifedom position) mediate uptake and benefits from livestock insurance in Northern Kenya. Intersectional analysis identified these hidden social groups (women in polygynous households) and highlighted the nuances in women's decision-making based on their marital status. The analysis of gender and household type shows the different ways that women may benefit or be disadvantaged within complex intrahousehold relationships. In monogamous households, women often participate in decisions about uptake, however in polygynous households women's participation is subject to intrahousehold power dynamics.

Insurance may collectively protect all household members' animals or a specific household member's livestock. The policyholder may purchase insurance for others, but pay-outs will go to the policyholders. In fewer cases, people in the study area talk of decisions to take one policy for the entire household (a pooled policy), in the name of whoever pays and can meet the other requirements (e.g., bank account). A husband's inclusion of all wives' input is socially valued, even if it is not always practiced. One male focus group discussant in Laisamis explained that *"It is difficult to make a decision because the families are always in disagreement when deciding on any decision"*. However, another man continued, *"The man of the house has to consult with all the wives before making any decision because if he does not, there will always be quarrels between the wives"* <*FGD\\ V1LAFG2_M6*>.

Within a household, some women can be excluded, mediated by husbands and dependent upon co-wife relationships. Samburu and Borana women described this exclusion, when husbands in polygynous households select a favorite wife(s). While this may be the last, and usually the youngest wife, this is not always so. The favorite wives usually benefit more from insurance coverage. Husbands may not even tell the other wives about the policy.

In Laisamis, women said that, "for polygynous families, the husband will use his income to buy insurance for the wife he loves the most." Some added that, "Mostly he will buy for the last wife." While "There are husbands who will buy the insurance for all his wives, and he will pay for them equally, at other times he will secretly buy insurance only for the loved wife <FGD\\V1LAFG2_F4>.

In Marsabit Central, women similarly reflected upon the practice of "good husbands" to pay for all wives. However a husband may choose who to buy insurance for based on her livestock activities.

"In a polygynous household, the husband can decide to consult a wife who is more involved with things about livestock," for instance. Or, "The husband might consult the wife he likes most." Another woman continued, "For the husbands who treat their wives equally, they will make decisions with them." In other case consultations are inclusive, however husbands may not provide money to buy insurance.

"The husband informs all the wives and each one will decide whether they want to buy, although the husband can decide to pay for one of them." <FGD\\ V2MCFG2_F5>.

Livestock insurance registration policies can reinforce such disadvantages because policies are under one name, but may cover multiple household members' animals. Livestock ownership is strongly gendered and women's claims to livestock assets vary among the Samburu, Borana and Gabra. Because of local conditions and what appears to be more flexible and changing gender norms and practices, Samburu women own livestock, especially sheep and goats.

9.3.4.2   Theme 2: Intersectional Analysis Helps Identify How External Shocks, in Concert with Development Processes, Interact with Intersecting Social Factors to Influence Livestock Insurance Decision-Making and Uptake Outcomes

Key Finding Livelihood precarity in Marsabit is exacerbated by climate change and community conflict. Intersectional analysis sheds light on coping strategies such as male outmigration and shifts in women's and men's income earning opportunities and expectations about who should earn. These experiences impact women's agency in decision-making related to livestock insurance in Northern Kenya. Women (and men) of different ethnicities and in monogamous v. polygynous households experience these processes differently. E.g., women become de-facto heads of household in some cases due to male outmigration while in others, despite men's outmigration, women still lack authority in decision-making.

Conflict between ethnic groups and climate change have compounded challenges for men and women, affecting livelihoods in significant ways. Women in Marsabit Central explained:

> There were clashes from 2017-2022 between the Gabra and Borana. Our livestock was stolen, houses were burned and many people lost their lives. But 2022 has been worse because we have lost almost all our livestock to drought. Many people are becoming depressed because they have lost all their livestock and they do not know how their families will survive. The few that were remaining are still dying, because of the lack of vegetation and some of us do not have money to buy hay V2_MC\\FGD\\V2MCFG2_F5>.

In response to such shocks male outmigration has increased. Men with livestock migrate in search of pasture and alternative income-generating activities. Men historically controlled income from livestock sales and have lost their main income source because of low market prices and multiple shocks that have killed livestock. In response, women, notably in Laisamis where urbanization is occurring, are stepping into more income-generation activities. This shift reflects increases in women's economic agency coupled with recent burdens to support households. Samburu women, primarily in monogamous households, occasionally purchase insurance without first seeking approval from husbands. In contrast, Borana and Gabra women reported seeking husbands' approval over mobile phones when men have migrated.

In Laisamis, women described how "For monogamous families, if the wife has money, she can decide to buy the insurance secretly without informing the husband." Another woman explained that this is linked to women's more recent activities earning income: "For now, most households depend on women's income for household needs, because men always depend on livestock for income and most of them have died due to drought." Another woman reflected that "Women always have different ways to get money" <FGD\\V1LAFG2_F4>.

These shocks affect livestock assets, especially those of women, because of intersecting social axes that exclude women from registering as policyholders. Women's purchase of insurance can secure their livestock assets in the face of multiple compounding factors.

9.3.4.3   Theme 3: Intersectional Analysis Reveals the Power Dynamics Embedded in Local Social Structures and Institutions that Shape Livestock Insurance Decision-Making, and How Agency Intersects with Local Institutional and Structural Constraints and Systems to Shape Livestock Insurance Decision-Making and Benefits

Key Finding Women from different ethnic groups and household types conform with, navigate and resist local socio-cultural norms related to livestock insurance. Samburu co-wives cooperate to secure insurance for their livestock. Women in both Laisamis and Bubisa, who are stepping into more income-earning activities may also buy insurance without first getting approval from husbands.

Cooperation between co-wives can improve their collective bargaining and power in their households. Women in Laisamis described how decisions are made about buying livestock insurance in polygynous households.

A woman recounted her personal experiences: "We will persuade the husband to buy the insurance for both of us, and he will do so. This depends on both wives being on good terms with each other; they can easily convince the husband to insure their livestock."

However, another explained, "If the co-wives are not in agreement, then the husband goes to each house to inform them about the insurance. Cooperation among co-wives is dependent upon a good relationship, which proves to also be strategic in terms of getting their livestock assets insured."

Women's agency is mediated by intrahousehold relations. Men often exercise authority in decision-making. However, through supportive co-wife relationships, women circumnavigate, resist, and overcome restrictive sociocultural norms pertaining to decision making about livestock and insurance.

9.3.5   Step 5: Using Intersectional Analysis to Inform AR4D Interventions

The case study findings reveal that certain social groups may be limited or unable to take up livestock insurance. Exclusion often occurs through intersections of ethnicity and relationships among primary adult members in polygynous households. These insights can be used to develop socially-inclusive AR4D interventions. A socially-inclusive IBLI scheme can address the needs of hidden and marginalized subgroups to ensure that they participate in livestock insurance. The analysis guided recommendations in product design, awareness, and availability, bundled services and strategies to target certain groups. Incorporating participatory approaches, such as codesigning and validating product features and dissemination channels with diverse community members, for example, is encouraged.

Product design features can improve different household member's potential to benefit from insurance. They can be flexible, such as offering household membership options that reflect different family members' multiple claims to livestock, and ensure fairer distribution of payouts to individuals. Policy member's household name would be used as opposed to an individual's name.

Hidden social groups may have less information about insurance products and subscription details, such as timing, location and product costs, which further hinders decision-making agency about whether to take up insurance. Education campaigns can raise awareness of the product and its value. Frequent broadcasting about insurance products and subscription periods through radio and SMS can enhance access to information.

Certain social groups require targeted outreach approaches. When males migrate with the herds, potential clients may be in remote satellite camps. Mobile subscription teams may travel to these locations for enrolment. Subscription periods can be planned when household members are more likely to be together, e.g., during rainy seasons or times when pasture is readily available.

Insurance, climate information and credit can be bundled to support uptake of services. Socially-transformative approaches can generate opportunities for peer learning and household dialogue. Radio campaigns on collective decision-making around insurance could be developed.

Insurance subsidies to marginalized groups could improve insurance access for certain individuals, households or communities. For example, to reach particular ethnic groups, a proxy such as geography could be used to target specific households.

Social change during times of crisis may occur rapidly and create opportunities for agency. Gendered power relations and intersections of gender with ethnicity are dynamic. In fragile contexts, crisis is compounded by multiple factors, such as climate change, economic precarity and conflict. As contexts change, research questions should ask: how do relations and agency in livestock and insurance decision-making, ownership, benefit, and risk distribution shift and afford more opportunities for certain individuals to purchase livestock insurance?

This case study can inform future applications of intersectionality in AR4D in several ways. First, it points to the evolving and fluid nature of gender relations and intersectional factors. Second, it illustrates the emerging changes in social and environmental contexts, especially the frequency and severity of climate shocks. These shocks require additional methods to track how intersectional analysis can inform insurance products to be better adapted to emerging and rapid change. These social changes include inter-marriage between ethnic groups, and the precarity and shifts away from livestock-based livelihoods.

9.4   Discussion

We now turn to how AR4D can deepen its understanding of intersectionality and its potential application to address multiple layers of inequalities and marginalization in agricultural research methods and practice.

Mixed Methods Enable Robust Insights on How Intersectional Inequalities Can Be Addressed Quantitative interactional analysis with large-n datasets typically explores which combinations of social characteristics emerge as the most relevant for a given research question. It identifies patterns in the social distribution of discrimination and opportunities. Qualitative intersectional analysis generally explores the mechanisms through which intersectional factors interact to shape respondents' lived experiences. It explains how and why particular interactions of social variables create specific outcomes. Within this case study, quantitative analysis of IBLI panel data informed the process for designing the qualitative components of the study.

Intersectional Analysis Unpacks the Complexities of Individual Lived Experiences Literature reviews and consultation with field staff are good starting points to identify relevant characteristics to explore. However, the study may overlook some issues as a result of these prioritization choices, because some characteristics are invisible to the majority.

Intersectional Analysis Could Help Develop Socially-Inclusive Interventions That Have an Impact at Scale The tension between uniqueness and scalability emerges from the fieldwork reported in this chapter and the researchers' approach to identifying axes that mattered most. Researchers may look for the most unique experiences or the most generalizable patterns of discrimination and opportunity. This depends at least in part on the research question at hand. Intersectional analysis highlights people's lived experiences along a continuum from disadvantage to privilege resulting from unique combinations of individual characteristics and axes of their interaction. Patterns in key axes that are most important to a given group may be relevant to some individuals only.

Focusing on diversity of lived experience may detract from efforts to design initiatives that can be widely scaled out. In the case examined here, impact of intersectionality at scale involves over 80,000 beneficiaries and intends to scale to 100,000 households across 16 counties in Kenya. Our work raises the question of how to identify uniqueness and diversity—intrinsic to intersectional approaches – while developing solutions that can achieve widespread societal impact. Intersectionality suggests that some solutions may not work for a large group. Yet, intersectional analysis may enable scaling strategies that address diverse needs rather than gloss them. Perhaps only some axes of disadvantage are widely shared. Only some will be relevant for the domains of interest. We propose that intersectional analysis can support the development of interventions that are capable of both social inclusion and impact at scale.

Limitations of the Method Highlight Selection Bias, Discussion of Sensitive Topics Like Ethnic Conflict, and Respondent Bias Limitations on the selection of intersection axes were primarily related to research methods that included the factors to focus on, sensitivity of topics, and possibilities of reporting normative behaviors in group discussions. We outline these three limitations and our efforts to address them below.

A deliberate and systematic focus on three axes was made, although participatory processes of both selecting and prioritizing certain axes would have been preferred to validate the selection. We may have missed other relevant factors. The analyses are thus not exhaustive, but given time and resource constraints these axes provided a first inquiry into intersecting factors. Additional research to explore other intersecting factors, like age, would help to understand exclusion of certain groups in insurance uptake.

A second limitation refers to sensitivity of discussing certain topics, especially where conflict is known to affect social groups along the lines of ethnicity. In this study, we sampled locations to solicit ethnically-specific gendered and cultural norms and practices related to livestock and insurance. Locations were a proxy for ethnicity, based on majority ethnic composition in each subcounty.

A third limitation refers to the quality of data solicited about socially unacceptable behaviors and practices. In group discussions there are risks that participants report normative behaviors as opposed to actual behaviors.

Polygyny (where one man has several wives) has been legal in Kenya since 2014 (Komen and Ling 2021). The codification of polygyny into law simply legitimated a pre-existing situation, with traditional systems of law historically recognizing polygyny. These dual systems influence tenure of land, livestock, and households. In Marsabit, polygyny is practiced, however tensions between members of polygynous households may exist, even if they are not discussed openly. We collected information about actual relations, harmonious or otherwise. We asked group participants about wider community trends, not their personal experiences. Discussing personal experience may compromise discussants' sense of privacy and result in inaccurate reporting of behaviors and practices. We also emphasized in the groups that were no right or wrong answers.

The IBLI Analysis Faced the Constraint of the Gender Norms Façade (Galiè and Farnworth 2019) Respondents providing answers that reflected idealized local gender norms rather than their actual experiences or opinions. Intersectional analysis focuses on how individual characteristics interact to shape differentiated individual experiences. However, such a focus may not increase people' ability to speak in public about their experiences. Close attention should be paid to how data are collected and interpreted.

Intersectional Approaches Are Particularly Relevant to AR4D in the Context of a Changing Climate The way individuals are affected by climate change depends on their positions in context-specific power structures based on social variables (Marty et al. 2022). People's responses to climate change are not only dependent on their social positions, but they are situated within other non-climate drivers that interact with climate change. Climate change adaptation is likely to transform livelihoods and social structures in ways that are difficult to predict. Research related to climate change would benefit from an explicitly intersectional approach to provide a more nuanced analysis and the tools to reveal different dimensions of identity and social differentiation in rapid and radical processes of socio-technical change (Mungai et al. 2017).

9.5 Conclusion

This chapter examines how an intersectional approach to AR4D can be operationalized at field-level. Intersectional approaches to AR4D offer a holistic way of understanding and addressing multiple layers of inequalities and sources of marginalization among smallholder farmers. The empirical case study on index-based livestock insurance (IBLI) in Northern Kenya demonstrated the value-addition of intersectionality as an analytical approach in AR4D by revealing hidden social groups and barriers to IBLI decision-making. It helps to identify how external shocks, in concert with development processes, interact with intersecting social factors to influence IBLI decision-making and outcomes. This conceptual lens enables an understanding of the power dynamics embedded in local social structures and institutions that shape IBLI decision-making. These findings can be applied to improve future IBLI interventions and inform gender-transformative approaches more broadly.

Given its novelty as an analytic approach in AR4D, there is scope for future applications of intersectionality. "The beauty of intersectional research is that its epistemology can always expand to consider a variety of socially constructed dimensions of difference that are salient in different contexts" (Misra et al. 2021, p. 17). As mentioned in the introduction, any attempt to operationalize intersectionality has a risk of instrumentalization –diminishing its value as an analytic concept towards transformative change. We caution against over-simplified approaches to "doing intersectionality" and suggest that future studies interested in exploring an intersectional approach stay true to the original framing of intersectionality as radical, transformative, and deeply political.

In the context of a changing climate, intersectional approaches can help reveal how social differentiation is implicated in rapid socio-technical change. Unless the rich nuance of intersectionality approaches are made available during the design stage of interventions, such projects are likely to be based upon assumptions and unlikely to mitigate vulnerability or close gender gaps. Many AR4D interventions are not able to bring lasting change because intersectionality approaches are not considered before the design of interventions. We encourage future researchers to document and share their best practices in intersectional analysis, and for research institutions such as One CGIAR to collate knowledge and support its own and other emergent communities of practice on intersectionality in AR4D.

Acknowledgements The empirical study on livestock insurance was supported by a grant from the BMGF grant and L&C initiative, with support from IBLI team members including Rupsha Banerjee and others.

References

Adaptation Fund (2022) Study on intersectional approaches to gender mainstreaming in adaptation-relevant interventions. 37th/38th Intersessional. https://www.adaptation-fund.org/wpcontent/uploads/2022/02/AF-Final-Version_clean16Feb2022.pdf

Ashik F, Voola A, Voola R, Carlson J, Wyllie J (2022) Advancing food well-being in poverty through intersectionality. Australas Mark J 30(4):1–10. https://doi.org/10.1177/1839334921998874

Axelrod M, Vona M, Colwell JN, Fakoya K, Salim SS, Webster DG, de la Torre-Castro M (2022) Understanding gender intersectionality for more robust ocean science. Earth Syst Gov 13:100148. https://doi.org/10.1016/j.esg.2022.100148

Badstue L, Petesch P, Farnworth CR, Roeven L, Hailemariam M (2020) Women farmers and agricultural innovation: marital status and normative expectations in rural Ethiopia. Sustain For 12(23):9847. https://doi.org/10.3390/su12239847

Bullock R, Crane T (2021) Young women's and men's opportunity spaces in dairy intensification in Kenya. Rural Sociol 86(4):777–808. https://doi.org/10.1111/ruso.12385

Calkin S (2015) Feminism, interrupted? Gender and development in the era of 'smart economics'. Prog Dev Stud 15(4):295–307. https://doi.org/10.1177/1464993415592737

Chant S, Sweetman C (2012) Fixing women or fixing the world? 'Smart economics', efficiency approaches, and gender equality in development. Gend Dev 20(3):517–529. https://doi.org/10.1080/13552074.2012.731812

Cho S, Crenshaw KW, McCall L (2013) Toward a field of intersectionality studies: theory, applications, and praxis. Signs J Women Cult Soc 38:785–810. https://doi.org/10.1086/669608

Choo HY, Ferree MM (2010) Practicing intersectionality in sociological research: a critical analysis of inclusions, interactions, and institutions in the study of inequalities. Sociol Theory 28(2):129–149. https://doi.org/10.1111/j.1467-9558.2010.01370.x

Clement F, Buisson M-C, Leder S, Balasubramanya S, Saikia P, Bastakoti R, Karki E, Koppen B (2019) From women's empowerment to food security: revisiting global discourses through a cross-country analysis. Glob Food Sec 23:160–172. https://doi.org/10.1016/jgfs.2019.05.003

Colfer CJP, Sijapati Basnett B, Ihalainen M (2018) Making sense of 'intersectionality': a manual for lovers of people and forests. Center for International Forestry Research (CIFOR). https://www.cifor.org/knowledge/publication/6793/

Collins PH (2002) Black feminist thought: knowledge, consciousness, and the politics of empowerment. Routledge, London

Cornwall A, Brock K (2005) What do buzzwords do for development policy? A critical look at 'participation', 'empowerment' and 'poverty reduction'. Third World Q 26(7):1043–1060. https://doi.org/10.1080/01436590500235603

Crenshaw K (1989) Demarginalizing the intersection of race and sex: a black feminist critique of antidiscrimination doctrine, feminist theory and antiracist politics. Univ Chic Leg Forum 140:139–167

Davis K (2008) Intersectionality as buzzword: a sociology of science perspective on what makes a feminist theory successful. Fem Theory 9(1):67–85. https://doi.org/10.1177/146 4700108086364

FAO (2023) Practical guide for the incorporation of the intersectionality approach in sustainable rural development programmes and projects. Food and Agriculture Organization, Santiago. https://doi.org/10.4060/cc2823en

Farhall K, Rickards L (2021) The "gender agenda" in agriculture for development and its (lack of) alignment with feminist scholarship. Front Sustain Food Syst 5:1–15. https://doi.org/10.3389/fsufs.2021.573424

Farnworth C, Ravichandran T, Galiè A (2023) Empowering women across gender and caste in a women's dairy cooperative in India. Front Sustain Food Syst 7:1123802. https://doi.org/10.3389/fsufs.2023.1123802

Ferguson CE (2021) A rising tide does not lift all boats: intersectional analysis reveals inequitable impacts of the seafood trade in fishing communities. Front Mar Sci 8:625389. https://doi.org/10.3389/fmars.2021.625389

Fisher M, Carr ER (2015) The influence of gendered roles and responsibilities on the adoption of technologies that mitigate drought risk: the case of drought-tolerant maize seed in Eastern Uganda. Glob Environ Chang 35:82–92. https://doi.org/10.1016/j.gloenvcha.2015.08.009

Flintan F (2021) Pastoral women, tenure, and governance. PIM Flagship Brief. International Food Policy Research Institute (IFPRI), Washington, DC. https://doi.org/10.2499/p15738coll2.134947

Galiè A, Farnworth CR (2019) Power through: a new concept in the empowerment discourse. Glob Food Sec 21:13–17. https://doi.org/10.1016/j.gfs.2019.07.001

Gewin V (2022) Decolonization should extend to collaborations, authorship and co-creation of knowledge. Nature 612:178. https://doi.org/10.1038/d41586-022-03822-1

Granovetter M (1985) Economic action and social structure: the problem of embeddedness. Am J Sociol 91(3):481–510. http://www.jstor.org/stable/2780199

Greene S, Pertaub D, McIvor S, Beauchamp E, Philippine S (2020) Understanding local climate priorities: applying a gender and generation focused planning tool in mainland Tanzania and Zanzibar. IIED. http://pubs.iied.org/10210IIED

Hancock AM (2007) When multiplication doesn't equal quick addition: examining intersectionality as a research paradigm. Perspect Polit 5(1):63–79. https://doi.org/10.1017/S1537592707070065

Hunting G, Hankivsky O (2020) Cautioning against the co-optation of intersectionality in gender mainstreaming. J Int Dev 32(3):430–436. https://doi.org/10.1002/jid.3462

Kaijser A, Kronsell A (2014) Climate change through the lens of intersectionality. Environ Polit 23(3):417–433. https://doi.org/10.1080/09644016.2013.835203

Kenya National Bureau of Statistics (KNBS) (2019) Kenya population and housing census. ISBN: 978-9966-102-09-6. Available: http://www.knbs.or.ke

Kojola E (2019) Indigeneity, gender and class in decision-making about risks from resource extraction. Environ Sociol 5(2):130–148. https://doi.org/10.1080/23251042.2018.1426090

Komen LJ, Ling R (2021) 'NO! We don't have a joint account': mobile telephony, mBanking, and gender inequality in the lives of married women in western rural Kenya. Inf Commun Soc 25(14). https://doi.org/10.1080/1369118X.2021.1927137

Kusakabe K, Sereyvath P (2015) Women fish border traders in Cambodia: intersectionality and gender analysis. In: Lund R, Doneys P, Resurrección BP (eds) Gender entanglements: revisiting gender in a rapidly changing Asia. NIAS Press, Copenhagen

Leder S, Clement F, Karki E (2017) Reframing women's empowerment in water security programmes in Western Nepal. Gend Dev 25(2):235–251. https://doi.org/10.1080/1355207 4.2017.1335452

Lokot M, Avakyan Y (2020) Intersectionality as a lens to the COVID-19 pandemic: implications for sexual and reproductive health in development and humanitarian contexts. Sex Reprod Health Matters 28(1):40–43. https://doi.org/10.1080/26410397.2020.1764748

Lotfata A, Munenzon D (2022) The interplay of intersectionality and vulnerability towards equitable resilience: learning from climate adaptation practices. In: The Palgrave encyclopedia of urban and regional futures. Springer, Cham, pp 1–16

Machado A, Good M (2022) Microinsurance and social protection: Ethiopia country case study. World Food Programme https://docs.wfp.org/api/documents/WFP-0000145513/download/

Marty E, Bullock R, Cashmore M, Crane T, Eriksen S (2022) Adapting to climate change among transitioning Maasai pastoralists in southern Kenya: an intersectional analysis of differentiated abilities to benefit from diversification processes. J Peasant Stud 50(1):1–26. https://doi.org/1 0.1080/03066150.2022.2121918

McCall L (2005) The complexity of intersectionality. Signs J Women Cult Soc 30(3):1771–1800. https://doi.org/10.1086/426800

McKune S, Serra R, Toure A (2021) Gender and intersectional analysis of livestock vaccine value chains in Kaffrine. Senegal PLOS One 16(7):e0252045. https://doi.org/10.1371/journal. Pone.0252045

Mello PA (2021) Qualitative comparative analysis: an introduction to research design and application. Georgetown University Press, Washington, DC. Online Appendix. Version 2.0

Misra J, Curington CV, Green VM (2021) Methods of intersectional research. Sociol Spectr 41(1):9–28. https://doi.org/10.1080/02732173.2020.1791772

Mitra A, Rao N (2019) Gender, water, and nutrition in India: an intersectional perspective. Water Altern 12(3):930–952

Mungai C, Opondo M, Outa G, Nelson V, Nyasimi M, Kimeli P (2017) Uptake of climate smart agriculture through a gendered intersectionality lens: experiences from western Kenya. In: Climate change management. Springer, pp 587–601. https://doi.org/10.1007/978-3-319-49520-0_36

Ng'endo M, Connor M (2022) One size does not fit all—addressing the complexity of food system sustainability. Front Sustain Food Syst 6:816936. https://doi.org/10.3389/fsufs.2022.816936

Ngum F, Bastiaensen J (2021) Intersectional perspective of strengthening climate change adaptation of Agrarian women in Cameroon. In: Leal Filho W (ed) African handbook of climate change adaptation. Springer, pp 2169–2191. https://doi.org/10.1007/978-3-030-45106-6_213

Nightingale AJ (2011) Bounding difference: intersectionality and the material production of gender, caste, class and environment in Nepal. Geoforum 42(2):153–162. https://iks.ukzn.ac.za/ sites/default/files/bounding%20difference.pdf

Oosthuizen A (2023) Intersectionality as a critical component of gender-transformative research. Human Sciences Research Council News and Events, 31 March 2023

Patil V (2013) From patriarchy to intersectionality: a transnational feminist assessment of how far we've really come. Signs J Women Cult Soc 38(4):847–867. https://doi.org/10.1086/669560

Ravera F, Martín-López B, Pascual U, Drucker A (2016) The diversity of gendered adaptation strategies to climate change of Indian farmers: a feminist intersectional approach. Ambio 45(Suppl 3):335–351. https://doi.org/10.1007/s13280-016-0833-2

Ravichandran T, Farnworth CR, Galiè A (2021) Empowering women in dairy cooperatives in Bihar and Telangana, India: a gender and caste analysis? AgriGender 6(1):27–42. https://doi. org/10.19268/JGAFS.612021.3

Ryder SS (2017) A bridge to challenging environmental inequality: intersectionality, environmental justice, and disaster vulnerability. Soc Thought Res Continuat Mid-American Rev Sociol 34:85–115

Seager J (2021) Gender and illegal wildlife trade: overlooked and underestimated. WWF. https://wwfint.awsassets.panda.org/downloads/gender_iwt_wwf_report_v9.pdf

Sverdlik A (2021) Gender and intersectionality in action research: taking stock, learning lessons and acting on opportunities. IIED Reflect & Act https://www.iied.org/20036iied

Tavenner K, Crane TA (2019) Beyond "women and youth": applying intersectionality in agricultural research for development. Outlook Agric 48(4):316–325. https://doi.org/10.1177/0030727019884334

Tavenner K, Crane TA (2022) Hitting the target and missing the point? On the risks of measuring women's empowerment in agricultural development. Agric Hum Values 39(3):1–9

Tavenner K, Crane TA, Bullock R, Galiè A (2022) Intersectionality in gender and agriculture: toward an applied research design. Gend Technol Dev 26(3):385–403. https://doi.org/10.1080/09718524.2022.2140383

Thompson-Hall M, Carr ER, Pascual U (2016) Enhancing and expanding intersectional research for climate change adaptation in agrarian settings. Ambio 45(Suppl 3):373–382. https://doi.org/10.1007/s13280-016-0827-0

Timu AG, Kramer B (2021) Gender-inclusive, -responsive and -transformative agricultural insurance: a literature review. CCAFS Working Paper no. 417. CGIAR Research Program on Climate Change, Agriculture and Food Security (CCAFS), Wageningen. https://hdl.handle.net/10568/117797

Van Aelst K, Holvoet N (2016) Intersections of gender and marital status in accessing climate change adaptation: evidence from rural Tanzania. World Dev 79:40–50. https://doi.org/10.1016/j.worlddev.2015.11.003

Vinyeta K, Powys Whyte K, Lynn K (2015) Climate change through an intersectional lens: gendered vulnerability and resilience in indigenous communities in the United States. General Technical Reports PNW-GTR-923 (72). U.S. Department of Agriculture, Forest Service, Pacific Northwest Research Station. https://www.fs.fed.us/pnw/pubs/pnw_gtr923.pdf

Whitley HT (2020) Power, privilege, and "playing in the dirt": an intersectional exploration of women's agricultural experiences in Pittsburgh, Pennsylvania. MSc Thesis Pennsylvania State University. https://etda.libraries.psu.edu/files/final_submissions/20398

Chapter 10
Feminist Research in Agriculture: Moving Beyond Gender-Transformative Approaches

Steven Cole, Surendran Rajaratnam, Millicent Liani, Deepa Joshi, Sahara Basnet, Meera Bisht, Meghajit Sharma Shijagurumayum, Mayank Jain, Prabhat Kumar, Kaitlin Fischer, Doris Puozaa, Alfredo Reyes, and Hazel Velasco

Abstract Feminist research approaches in agriculture are considerably underutilized. In this chapter, we suggest a few key reasons to help explain their lack of use in agriculture. We also provide background on what constitutes feminist research in agriculture through a review of the literature. Using a case study approach, we highlight the important and unique characteristics that define feminist research approaches in agriculture. The case studies provide examples of how researchers working in agriculture can gradually adopt key feminist research principles. We argue that to transform agrifood systems to be more inclusive, equitable, and sustainable, feminist approaches must be used in all research in agriculture. The chapter concludes by discussing what is needed to increase the use of feminist research approaches in agriculture, recognizing that resistance to change is inevitable and requires commitment at the top to spearhead efforts to institutionalize feminist approaches within agricultural research organizations.

S. Cole (✉) · M. Liani
International Institute of Tropical Agriculture, Dar es Salaam, Tanzania
e-mail: s.cole@cgiar.org

S. Rajaratnam
Universiti Kebangsaan Malaysia, Selangor, Malaysia

D. Joshi · S. Basnet · M. Bisht · M. S. Shijagurumayum
International Water Management Institute, Colombo, Sri Lanka

M. Jain · P. Kumar
SumArth, Bihar, India

K. Fischer · A. Reyes · H. Velasco
The Pennsylvania State University, University Park, PA, USA

D. Puozaa
Savanna Agricultural Research Institute of the Council for Scientific and Industrial Research, Tamale, Ghana

© The Author(s) 2025
J. Njuki et al. (eds.), *Gender, Power and Politics in Agriculture*,
https://doi.org/10.1007/978-3-031-60986-2_10

10.1 Introduction

An equitable and sustainable transformation of agrifood systems must embody the use of feminist approaches (Park et al. 2021). Farhall and Rickards (2021, pp. 1–2) argue that the use of feminist approaches entails tackling "forms of power and privilege within agricultural production and supply chains to include more diverse human voices and address structural issues … [which is critical] because unnuanced gendered approaches to development can exacerbate inequalities, re-entrench forms of difference, or marginalize women in new ways." While gender is clearly on the agriculture for development agenda, now more so than ever, feminist approaches in agricultural research remain significantly underutilized (Farhall and Rickards 2021).

Several reasons help explain the lack of use of feminist approaches in mainstream agricultural research. First are the epistemological and methodological differences between and within organizations that carry out gender-related or women-focused research in agriculture (see Feldman 2018) and their staff capacities to adopt and implement feminist approaches (Travis et al. 2021). For example, the Australia-based International Women's Development Agency (IWDA 2017) has a feminist research framework for use by its staff, which includes an approval process that staff must follow as they design and implement their research, analyze data, and communicate their findings for action. In contrast, some research and development organizations have gender strategies that guide, rather than mandate, researchers and practitioners on how to carry out sex-disaggregated analyses and integrate gender perspectives in their work.[1] Far fewer organizations carry out strategic gender research that prioritizes gender topics in agriculture. In recent years there has been a move away from research questions across different scientific disciplines that assume only men are farmers, agricultural managers, or decision-makers, as well as conclusions drawn from male-only samples while claiming universal or generalizable application (Feldman 2018), however research capacities within organizations to use feminist approaches are still low.

Second, there is a propensity for most agricultural research organizations to focus on short-term outcomes associated with their work, for example, when researchers from a given organization work with women to increase their access to and uptake of improved crop varieties for enhanced productivity and profitability. These outcomes are often achieved using a gender-responsive approach that develops innovations for women and men based on their practical gender needs rather than by setting up research processes to understand strategic gender needs and address the power differentials at household and other institutional levels that

---

[1] See the following examples for different gender strategies that support agricultural research and development: https://gender.cgiar.org/about-us/gender-strategies; https://www.jica.go.jp/english/our_work/thematic_issues/gender/c8h0vm0000f3jmj6-att/gender_mainstreaming_07.pdf;   and https://www.fhi360.org/sites/default/files/media/documents/FHI%20360_Gender%20Integration%20Framework_3.8%20%2528no%20photos%2529.pdf

exclude or subordinate women in agriculture and the broader society (Njuki et al. 2022). As such, women become targeted by researchers and practitioners as a means to increase economic, food and nutrition security (see Elias et al. 2021). The use of an instrumental approach to agricultural development, according to feminist schol- ars, is far more common than the use of an intrinsic approach that promotes gender equality as a goal in and of itself (see Cole et al. 2015; Farhall and Rickards 2021). The former approach, which focuses on individual capacity-building, can divert the focus away from addressing the causes of gender inequalities through collective mobilization (Farhall and Rickards 2021).

Third, there is a general resistance within the agriculture sector (but also within other sectors) to embrace gender equality or gender-aware approaches, let alone feminist principles, including when carrying out research and development work (EIGE 2016; Rao 2005; Kabeer 2007, 2016). The resistance towards feminism grows when it is viewed as gaining too much power or when feminists become suc- cessful at challenging patriarchal structures (Ikävalko and Kantola 2017). Individual and collective movements against patriarchy and the structures that maintain harm- ful practices within institutions are often challenged and can result in the creation of new counter movements that put hard-won rights at risk[2] (Shameem 2021). Resistance to feminism can (and often does) take the form of silence in response to practices that create and perpetuate gender inequalities (Ikävalko and Kantola 2017).

This chapter highlights the important and unique characteristics that define femi- nist research approaches in agriculture, by presenting four purposively-selected case studies. The case studies provide examples of how researchers working in agri- culture can gradually adopt key feminist research principles. While conducting feminist research in agriculture is challenging and requires significant commitment to people and place, we argue that to transform agrifood systems to be more inclu- sive, equitable, and sustainable, feminist approaches must be used in all research in agriculture.

The authors of this chapter all consider themselves feminists who use feminist principles in the research they conduct in agricultural contexts, with a strong desire to bring about transformative change from the work we do. Case study authors are women and men from diverse countries in the Americas, Africa, and Asia, with varied educational backgrounds, and development and research experiences work- ing on gender issues within their organizations. The authors acknowledge here that such experiences and training in equally diverse theoretical perspectives shaped how we framed the four case studies.

---

[2] We acknowledge the valuable contribution of an anonymous reviewer of the book chapter who raised this point.

10.2   What Is Feminist Research in Agriculture?

Feminist values must underpin all aspects of research efforts to contribute meaningfully to women's rights and the achievement of gender equality (Jenkins et al. 2019) as well as to transform agrifood systems (Park et al. 2021). Accordingly, there are several frameworks or lists of principles to mandate or guide the design and implementation of feminist research. The framework developed by the IWDA (2017) is useful for an understanding of the mandatory components of doing rigorous feminist research, highlighting four key components: (1) building feminist knowledge of women's lives, (2) accountability for how research is conducted, (3) commitment to ethical collaboration, and (4) having a transformative impact on the causes of gender inequality. We use this framework to help structure the literature reviewed in this section on what constitutes feminist research in agriculture and also the case studies we present in the next section.

Feminist research differs from gender research in that it aims to examine the diversity of women's experiences and how gender norms and power relations create inequalities between women and men (IWDA 2017; Kiguwa 2019). Podems (2010) argues that feminist research examines why gender differences exist and challenges women's subordinate position while acknowledging the multiple variations between women that shape their experiences with oppression in different ways (see also Jenkins et al. 2019). Others stress that examining the impact of intersectionality (versus intersecting identities) on women's lives is a salient feature of doing good feminist research (IWDA 2017; Mullinax et al. 2018; Kiguwa 2019) and requires that researchers consider how systems of inequality based on sex and gender identity, ethnicity, skin color, age, sexual orientation, geographic location, colonial history, among many other forms of discrimination and oppression, intersect to create unique experiences, dynamics, and outcomes.[3] According to Kiguwa (2019, p. 227), intersectionality is "a core political tool of feminism" and the scholarship on intersectionality is quite diverse.

Feminist research prioritizes ethical approaches by adopting the precautionary principle of "do no harm" (IWDA 2017; Mullinax et al. 2018), which requires that the research does not create any additional risk due to people's involvement in the research. While the notion of a universal feminist research ethics is unreasonable given a multitude of feminisms and the use of different methods by feminist researchers (Kingston 2020; Kiguwa 2019), key ethical standards would include, for example, ensuring confidentiality and safety, informed consent, and respect for all research participants and research team members.

Feminist researchers in agriculture use diverse methods to examine power relations and patriarchy and the impacts they have on creating and perpetuating gender inequalities. By using multiple methods, feminist researchers can understand and present diverse worldviews of women in different ways (Kiguwa 2019). Tickamyer (2020) notes that using a feminist research approach does not necessarily mean only

---

[3] https://www.intersectionaljustice.org/what-is-intersectionality

using qualitative research methods, but rather using both qualitative and quantitative tools (see Jenkins et al. 2019) to address the research and societal problems, while also taking into consideration the research setting. Kiguwa (2019) notes, however, that past writings on this topic suggest that the values of quantitative tools and methods are, in themselves, problematic for failing to make sense of the social world and lived realities of many women.

Historically, feminist researchers use participatory research methods that aim to identify discriminatory norms and unequal power relations and determine suitable actions to address these underlying causes of gender inequalities (IWDA 2017; Jenkins et al. 2019; Njuki et al. 2022). A key characteristic of participatory feminist research is the iterative, circular, flexible, and dynamic nature, which assists in disentangling social and gender inequalities and empowering those who have been silenced (Mullinax et al. 2018). Participatory approaches cultivated by feminist scholars often emphasize critical reflexivity, the inclusion of disenfranchised voices, and dialogical problem-solving. As such, the researcher is not regarded as an objective expert, but rather aims to set up each stage of the research process to encourage the active participation of women, develop their capacities, and enable them to feel empowered by the process. Feminist research embodies the notion that the research being conducted is "for and with women" rather than conducting research "on women" (IWDA 2017, p. 15; see also Leung et al. 2019).

Feminist researchers pay particular attention to the fact that they enter the research process with a set of values that must be questioned throughout as it influences how the research is conducted, interpreted, and communicated (IWDA 2017; Jenkins et al. 2019). Feminist research explicitly recognizes the power dynamics involved when conducting research with women, and therefore, demands that researchers remain cognizant and reflexive about these dynamics of the research relationship throughout the research. Researchers must think critically about their relationships with the social world and their understandings of their experiences (Webster et al. 2014). Being reflexive encourages researchers to be honest with themselves about their motivations for participating in a particular research project as well as about their positionality when engaging in research (Manning 2018). This is particularly relevant in the context of research that examines multiple axes of difference. Critical research examines power relationships, explores the complexities of positionality and representation, and questions the researcher's position as (re) presenter of the participants (Ozkazanc-Pan 2012). Researchers must go beyond noting personal beliefs and assumptions and how they affect interactions with people.

Feminist research aims to move our understanding of women's lives in new directions by researching neglected issues, and in particular, the root causes and consequences of gender inequality, and ensuring the research is action-oriented, so that discriminatory norms and unequal power relations are transformed for greater gender equality (IWDA 2017; Kiguwa 2019). Prior work to empower women has often failed because of little or no regard for "the intersectionality of discrimination against women" and "the deeply ingrained nature of gender inequality at a structural and political level" (Mullinax et al. 2018, p. 4). Feminist research can help bring about transformative change at multiple systemic levels, from the individual to the

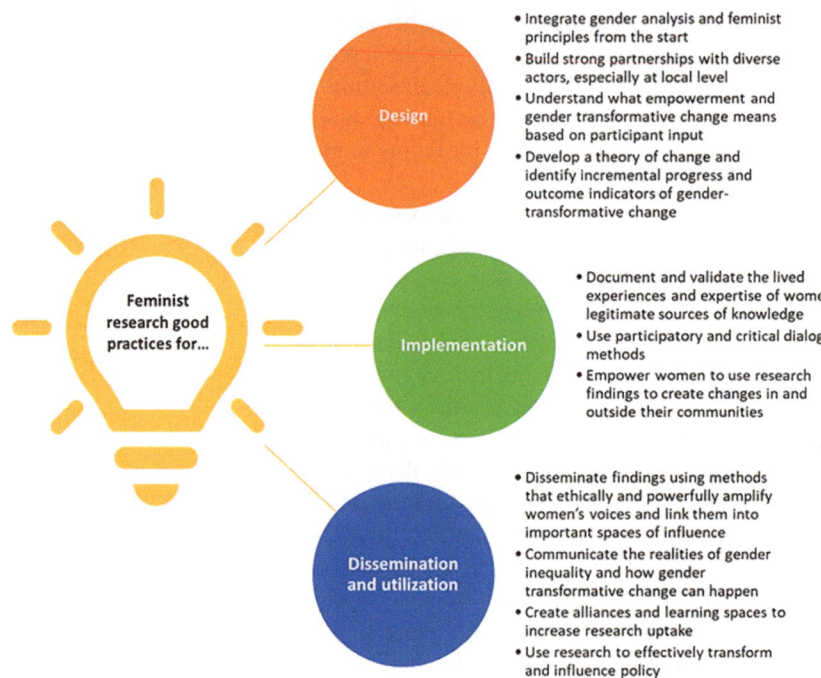

Fig. 10.1 Good practices for effective feminist research

organizational up to the societal level, and within movements and through partnerships, and in how research and knowledge is produced (IWDA 2017). Moreover, feminist research aims to inform the design and promotion of responsible technologies and influence policy, practice, and programming to help create an enabling environment for gender-transformative change (Cadesky 2020).

Based on these principles, Mullinax et al. (2018, p. 6) have summarized the good practices for effective feminist research design, implementation, and dissemination and use (Fig. 10.1). Many of these good practices are highlighted in the case studies presented in the next section.

10.3   Good Practice Case Studies Using Feminist Research Approaches in Agriculture

Four case studies are presented in this section to showcase how feminist research approaches can be used in agricultural research. Researchers were selected to develop their case studies based on prior knowledge that their agricultural research embodied some of the key feminist research principles detailed in the section above. The lead authors of this chapter asked researchers to respond to a prompt, or a series of guiding questions, that forms the structure of each case study from the perspectives of those who carried out the research. The lead authors also developed a case study to pilot the prompt before sharing it with others.

The guiding questions included in the prompt (see Appendix) were developed after reviewing the literature on what constitutes a feminist research approach. The overall structure of the prompt was informed by the four component parts of the IWDA (2017) framework on doing rigorous feminist research. The other literature reviewed helped us to include specific guiding questions under each of the four component parts of the prompt. While unintended when designing the prompt, it is now apparent that the guiding questions in Appendix are useful in helping other researchers to design, implement, and monitor and evaluate their feminist research in agriculture.

10.3.1   Case 1: Gender-Transformative Research in the Barotse Floodplain of Western Province, Zambia

Steven Cole and Surendran Rajaratnam

We carried out gender-transformative research (Cole et al. 2014a) from 2013 to 2018 in the Barotse Floodplain of Western Province, Zambia. The research was part of a larger CGIAR Research Program on Aquatic Agricultural Systems (CRP AAS) and was informed by feminist research principles (see Kantor and Apgar 2013; Kantor 2013). Commitment to people and place was a mainstay throughout the research. While the research used mixed methods in a range of smaller research initiatives on different topics, participatory action research (PAR) cut across this work to ensure that it helped address the challenges faced by women and men who depend on the floodplain for livelihood security.

The research began with an understanding of the lived experiences of women and men living in 10 large communities in the Barotse Floodplain. The researchers carried out a mixed-methods social and gender analysis, with a strong focus on understanding the norms and power relations that create gender inequalities in the floodplain, from the perspectives of women and men who shaped and were shaped by such inequality. A wide range of science and communication outputs were developed during this initial phase of the research (see Cole et al. 2015; Rajaratnam et al. 2015, 2016; Dierksmeier et al. 2015) that later helped design several interventions that aimed to tackle the root causes of gender inequalities in the floodplain.

The gender-transformative research worked with women and men in the floodplain to understand how unequal power relations created advantages for some and disadvantages for others, while also creating sub-optimal development outcomes at household, community, and higher levels. The research used different qualitative tools to understand how certain norms and practices have changed or remained constant over time (Dierksmeier et al. 2015; Rajaratnam et al. 2015). Such perspectives can enable women and men to see that norms and attitudes are mutable over time, and thus, transformative change is possible when women and men work together to achieve positive development outcomes.

While the research primarily included rural and resource-poor women and men, it captured a range of different socio-demographic and economic characteristics of research participants to include the experiences of diverse sub-groups of women and men. The researchers used an intersectional lens to understand and depict (via a well-being ladder) how multiple axes of identities intersect and interact to impact on women's and men's lives (see Rajaratnam et al. 2015, pp. 34–40). The research listened to women and men from different backgrounds from many communities throughout the floodplain.

The researchers studied both women/femininities and men/masculinities throughout the five years of research in the Barotse Floodplain. The work on rural masculinities and their impacts on disadvantages for women and other household members (Cole et al. 2015) helped to develop gender-transformative interventions to tackle restrictive norms and power relations.

The research was based on a gender-transformative theory of change (Cole et al. 2014a, b) to assess how gender-transformative change occurs, including initial changes (McDougall et al. 2015) and across different social change interventions. The researchers assessed the changes in women's empowerment outcomes and gender equal attitudes within a post-harvest fish loss intervention (Cole et al. 2018; Cole et al. 2020) and decision-making powers within a savings group intervention (Cole et al. 2021).

Over the five years, the project designed and implemented the research with various social and biophysical scientists, extension officers, value chain actors, and community members. While all publications included research team members with various educational backgrounds, e.g., from WorldFish, Department of Fisheries (national, regional, and district levels), and University of Zambia, the researchers failed to bring research participants onboard as co-authors, yet did acknowledge their contributions throughout the research.

The research understood the restrictive norms and power relations that create gender inequalities in the floodplain and set up iterative cycles of critical reflection, action planning, doing, and learning. By design, the research did not exploit or accommodate existing norms that restrict women from engaging in and benefiting from fishery-related activities. For example, solutions to the challenges women and men faced adhering to discriminatory norms came from the research participants themselves and not the outside researchers. While this cannot ensure that research keeps all participants from additional harm throughout the research process, it did ensure that those involved in the research were willing to try new ways of thinking and being that did not spark backlash or negative outcomes.

The PAR that utilized many feminist research principles was used during different stages of the research and across its topics over the five years. For example, the action research on post-harvest fish losses first understood how fish loss and waste in the floodplain was gendered, and subsequently set up a participatory process to select and modify improved processing technologies to fit women's needs and preferences. The researchers also implemented a social change intervention using drama skits and critical reflection and action planning sessions on the restrictive norms and power relations that create fish loss and waste, among other issues. The researchers set up a monitoring and evaluation system to determine what changes in gender relations were happening and how (Cole et al. 2020, 2021).

Research findings throughout the five-years were disseminated and validated using strategies that ranged from feedback from research participants after trying out actions that were formulated during action planning at group level (Cole et al. 2018, 2020, 2021) to validation of the baseline and benchmarking data (Rajaratnam et al. 2015) to large stakeholder meetings to determine additional ways of supporting inclusive and sustainable value chain development (Kaminski and Cole 2018). Traditional ways of disseminating research findings were also used, including in the publications cited in this case study and at end-of-project stakeholder workshops.

During the five years, the research created alliances and learning spaces within Zambia and elsewhere to increase research use. This body of research has created the evidence that the use of a gender-transformative approach, informed by feminist research principles, can work and helps facilitate the empowerment of women and men, while bringing out additional positive development outcomes. For more information see Wong et al. (2019), McDougall et al. (2021), and the ongoing European Union-Rome Based Agencies Joint Programme on Gender-Transformative Approaches for Food Security, Improved Nutrition and Sustainable Agriculture (see FAO 2022).

The research used a two-fold approach to convince audiences of the realities of gender inequality and to communicate how gender-transformative change can happen: (1) PAR with stakeholders working at community, district, regional, national and international levels; and (2) traditional science and communication outputs. While changing formal policies were not an explicit focus of the research, policy and decision makers within organizations operating at different scales were engaged with the research at different times. And although uncompleted due to funding constraints, the researchers carried out an institutional analysis to understand how organizing committees at the district level can be part of the transformative change process (Kato-Wallace et al. 2016).

The research team was reflexive and introspective at all stages of the five-year research. A range of workshops and meetings were held to critically reflect on and plan the research. Gender-transformative theory of change workshops were incredibly useful in this regard, especially when identifying how institutional change must happen before or while implementing transformative change outside one's institution. During capacity-development workshops on how to implement gender-transformative approaches, the gender equal attitudes of workshop participants were assessed to determine whether they were changing. At another workshop, the researchers set up interesting role plays that helped question the mindsets of different research team members. Gender research capacities were also developed during the five years to enable research team members to carry out gender research in the future.

During various stages of the research, the team interrogated the power dynamics associated with the research relationship either using PAR with team members and research participants or during research planning meetings when all team members were given the chance to input and shape the direction of the research and its outputs.

The research intervened primarily at individual, household, and community levels, with further engagement at the organizational level (e.g., through gender capacity development and institutional change efforts). Change occurred at the individual and relational levels, as evidenced by the monitoring and evaluation and via publications. Team members learned how to do gender-transformative research with multiple stakeholders and working in a complex socio-ecological system, which continued to yield results as team members tried to facilitate gender-transformative change in agriculture at scale.

While the researchers embraced an intersectional lens during the research, not all analyses and write ups showcased the intersectional approach used. They acknowledged that at times they wished to disaggregate their analysis further according to ethnic group or age, but this proved challenging with the quantitative data given the small sample sizes at these disaggregated levels. A focus on youth was also limited during the research.

10.3.2   Case 2: The Women in Agriculture Network: The Role of the Horticulture Value Chain in Empowering Women and Indigenous Populations in Honduras' Dry Corridor

Janelle Larson, Paige Castellanos, Leif Jensen, Carolyn Sachs, Arie Sanders, Alfredo Reyes, and Hazel Velasco

The Women in Agriculture Network (WAgN) research project in Honduras was a five-year collaborative effort between The Pennsylvania State University (Penn State) and Zamorano University. The research project was part of the USAID Feed the Future Innovation Lab for Horticulture at the University of California, Davis (UC Davis). The project explored whether the horticulture value chains could be a mechanism for empowering women and indigenous populations in Honduras' Dry Corridor region. From 2015–2019 the project studied the most critical barriers to successfully including women in horticulture value chains. The WAgN-Honduras team implemented mixed feminist research methods to investigate whether smallholder women farmers' participation in the horticulture value chains could positively impact their food security and access to extension services. The project employed in-depth interviews, focus group discussions, surveys, social network analysis, and participatory extension methods using a gender-transformative design. See references in Larson et al. (2019) and Sanders (2021).

We began the research by identifying key stakeholders in the Honduran Dry Corridor region and establishing positive working relationships with them. Stakeholders in the project included public and private agricultural development institutions, NGOs, farmers, and women's associations. Through stakeholder engagement, the research team gained a rich understanding of gendered roles, division of labor, and inequalities in the region. Key findings of the research included:

- Despite multiple attempts to involve women in the agricultural value chain, efforts typically entailed using a top-down approach and focused primarily on the economic aspects of women's empowerment.
- Despite having a women's inclusion policy, most existing farmer's associations were dominated by men. Often, women were included to fulfil the requirements of donors and obtain funding. As a result, women were underrepresented in group membership and had limited access to its benefits, such as training.
- Women's organizations were more likely to prioritize non-commercial crops, indigenous knowledge and culture, and food sovereignty.

Based on the initial results, the WAgN-Honduras team included gender-transformative elements in the design and implementation of two farmer field schools (FFS). Due to the male bias of most farmers' associations in the region, the team implemented the two FFS with an indigenous women's association to ensure that women were meaningfully included. In the rest of this case study, we describe our experience with transformative FFS to provide a detailed example of our feminist research approach.

To assess transformative change, the team conducted ex-ante and ex-post semi-structured interviews and surveys. Participants were asked questions about their aspirations to produce high-value vegetable crops, their views on gender equality, and their expectations and reflections on the experience.

Two FFS were implemented with smallholder female and male farmers living in the rural region of Intibucá, Honduras. In the results of the initial data collection of the research project, the team identified a gender gap in access to agricultural assets and information. Thus, the purpose of the FFS was to explore suitability of the farmer field school to facilitate women's human capital formation and access to assets.

The FFS approach has traditionally focused on the diffusion of knowledge-intensive integrated agricultural practices (Godtland et al. 2004). More recently, it has also been used to promote community-level discussions related to gender, nutrition, empowerment, and gender-based violence (Davis et al. 2012). Despite differing findings regarding the impact of FFS on participants, research has confirmed that intersectional factors such as gender, race, class, and education can have meaningful implications for the program's outcomes (Choudhury and Castellanos 2020). Thus, the primary objective of the FFS was to address the gender gap in agricultural knowledge by implementing participatory extension. Implementing the FFS was an attempt to mitigate the burden of participation for frequently underserved groups— indigenous women—by incorporating a gender-transformative framework into the design, implementation, and evaluation.

The FFS was designed and carried out as a tripartite collaboration between: (1) The WAgN-Honduras team, consisted of Penn State and Zamorano faculty and two research assistants from different disciplines. The team was responsible for the oversight of the FFS curriculum, gender-responsive framework implementation, data collection, and evaluation of the FFS method. The WAgN-Honduras team provided expertise in participatory research, gender, and agriculture, and already had a long-term relationship with the partner association; (2) Zamorano's Horticulture Innovation Lab team was responsible for implementing the agricultural section of the FFS curriculum. The Horticulture Innovation team had previous experience in FFS with smallholders in Honduras' Dry Corridor; and (3) the *Asociación de Mujeres Intibucanas Renovadas* (Association of Renewed Women of Intibucá— AMIR) staff who identified the research participants and provided feedback during the whole process to adapt the curriculum and feminist framework to the association's interests.

The FFS were developed in two different communities in the department of Intibucá, Honduras. Intibucá is in the Honduran Dry Corridor and has some of the highest rates of poverty and food insecurity in Honduras. The communities and the participants were selected with the support of the partner indigenous women's association, AMIR. Interested participants were selected based on the following criteria: they should be (a) smallholder farmers, (b) at least 18 years of age, and (c) a member of AMIR. Two groups of 25–35 people each were formed with the association's assistance. One FFS group consisted of women only (n = 34) and the second group was mixed (20 women and seven men). AMIR also requested that each group include participants from different nearby communities to increase cross-community partnerships. The final sample consisted primarily of indigenous Lenca women, and some Lenca men, and the sample captured a range of ages, family composition, economic characteristics, and levels of education. Our results were compared and contrasted based on participants' socio-demographic profiles, allowing us to conduct an intersectional analysis. In a feminist critical approach, all experience is considered intersectional, so there is no universal, homogenous experience of gender, race, class, or sexual orientation (Allen 2022).

The project was focused on studying indigenous rural women's realities and barriers to participate in agriculture. We also studied gender dynamics and how these women performed gendered tasks. We examined the women's expectations and how these are constructed vis-à-vis their male counterparts. By looking at gender dynamics inside and outside the household and smallholders' associations and society overall, the research team was able to identify and discuss with research participants how these divisions impacted the access and wellbeing of individuals from an intersectional lens. For example, during one of the discussions during the FFS the team explored how the construction of gender roles was intertwined with other aspects such as age, marital status, land ownership and its repercussions for different individuals to access resources and pursue interests.

The research was based on a theory of change. It hypothesized that by increasing agricultural knowledge and reducing women's burdens to participate in agricultural extension programs, women's productive agency would increase, thus improving food security and overall wellbeing for their families. By examining participants' perceptions of the benefits and drawbacks of gender-transformative-participatory projects, this research helped to identify specific methodological or logistical limitations that may cause gaps between the project's aims, its immediate results, and its lasting effects on agrarian societies in the Global South.

The analysis of the data was conducted by the WAgN-Honduras team with constant feedback from the women's association. Some leaders from the women's association have participated as co-authors during oral presentations of the research results. This study was approved by the Internal Review Board (IRB) of Penn State University (STUDY00008275 and STUDY00017215). Informed consent was obtained through a verbal consent process before the start of the interviews, focus group discussions, and FFS sessions. To ensure transparency during the process and participants' comfort, all activities were conducted in Spanish.

During research activities, transportation costs were covered for the participants and childcare was provided at each meeting, as well as a nutritious meal for the participants and their families. In addition, with the help of the participants and the partner associations, the days and times for the research activities were established to improve participants' attendance. In developing logistics, the partner association played an essential role. In doing so, they helped determine what was less burdensome and what might be seen as a more compelling incentive for people to participate in the research. This included wages paid to cooks and nannies, as well as what kind of food and how much to provide.

Research findings were disseminated and validated in an iterative process, including with FFS participants and women's association leaders. The research design included multiple methods to ensure the robustness of the data such as pre- and post- semi structured interviews with participants, focus group discussions, observations, and short- and long-term follow-ups. The results were presented and discussed in multiple occasions during these visits with the participants, the women's association, and local and regional organizations working with smallholder farmers, as well as with other stakeholders. Results have been presented in academic conferences in Central America, the United States of America (USA), and Australia, and manuscripts are being prepared to publish the experience in peer-reviewed journals in English and Spanish.

The FFS were developed in alliance with scholars, agricultural extensionists, and an indigenous women's association (AMIR). One of the goals of this research project was to design and pilot a gender-transformative FFS that could be used as a model by development organizations in the region and beyond. So far, the women's association has been able to secure funding from three other international development organizations to conduct an adjusted version of the FFS that was part of this project. AMIR used their expertise gained in the design, evaluation and results of this project to adjust the FFS method and continue working on securing smallholders' access to knowledge and strengthen their communities and organization.

To convince audiences of gender inequality and explain how gender-transformative change can be achieved, the team used participatory research involving stakeholders at the community, regional, public and private levels, and traditional academic approaches.

By using FFS, and PAR, the researcher becomes a facilitator rather than an expert and an activist rather than an independent neutral scholar. At each step of the research, the WAgN-Honduras team held meetings with stakeholders and partners to ensure the research remained relevant to their interests, mainly centered on the needs and realities of women participants.

Feminist research approaches call on us to center power relationships during the whole research process. A key aspect of our ability to carry out the research was the care and effort we invested in developing rapport with stakeholders. The team tried as much as possible to design and adapt the research agenda with the goals of social equality, so the research could be of direct use to these stakeholders. Their expertise and knowledge of the region and smallholders' realities was constantly key in shaping the focus of the research as well as for refining the research methods and working directly with research participants.

In general, the team believes the research had a positive, transformative impact at the individual and organizational level. In a post-evaluation, all participants mentioned that learning more about the methods and techniques of agricultural production was extremely rewarding. Several farmers emphasized the importance of learning, even when the topics were not new to them. Most explained that it was positive to have a collaborative space with other farmers and facilitators to reaffirm what they have learned empirically on their own and to ask questions and listen to different ways of farming.

Participants appreciated the opportunity to socialize with others in their community and with people from nearby villages, with whom they did not often have the chance to interact. This gave them the opportunity to coordinate other social activities within the community and association. The status they gained at the community level led to them being asked for advice. They also mentioned the relationships built during the FFS helped them to connect with others outside their communities. Other participants mentioned they started trading seeds among each other, and some even set up communal school gardens.

There were several limitations beyond the scope of this case study, but they are worth considering in the design of future feminist-oriented research and extension projects. First, the oversampling of women provided rich data about women's particular experiences but limited information about men's experiences. Second, it was not possible to determine how or if the gender and nutrition-related discussions during the FFS sparked other conversations in the household or community. In describing examples during the evaluations, participants referred more to the women's association's ongoing efforts to create more equitable communities for women than to the FFS per se. In addition, lack of time and resources limited the research team's ability to capture farmers' experiences in-depth and continuously throughout the sessions.

10.3.3 Case 3: Climate Vulnerabilities and Resilience of Marginalized Groups in Bihar, India

Deepa Joshi, Sahara Basnet, Meera Bisht, Meghajit Sharma Shijagurumayum, Mayank Jain, and Prabhat Kumar

Climate impacts amplify agrarian distress in Bihar, India, which has been historically shaped by deep-rooted, social, economic, and political challenges. The Doing Science with Society (DSWS) project was funded by the CGIAR GENDER Impact Platform and was carried out by the International Water Management Institute (IWMI) in collaboration with two implementation partners, SumArth and Cynefin Co. The project aimed to understand climate vulnerabilities and resilience of marginalized groups, particularly women, in the Gaya district of Bihar. The focus was to unpack the multidimensional nature of inequality, and to understand how climate challenges shape and reinforce gendered vulnerabilities in agriculture. The project also focused on unpacking deep-rooted values and biases in institutions engaged in planning, designing, and implementing agricultural interventions.

The main implementation partner in Gaya was SumArth, a farmer-producer collective of over 13,000 farmers (60% women) working in seven Bihar districts to achieve reliable, profitable agriculture. Over 2000 SumArth members are landless, socially marginalized, and resource-poor, some of the estimated 374 million multidimensionally poor people in India, who lie outside the focus of macro-level policy interventions. Many in this group engage in peripheral agrarian practices in the informal sector and often rely on solidarity and reciprocity as key survival strategies.

In this research, we applied a transdisciplinary, ethnographic, digital tool called SenseMaker, which allows combining data analytics with personal, unique human stories and experiences. Case study data (597 stories) were collected from farmer end-users (382 women and 215 men) and 80 institutional stakeholders (51 women and 29 men). Individual and focus group discussions were also conducted with many of the same stakeholders.

Informed by a feminist political ecology lens, our focus in this research was:

1. Collating individual, embodied gendered experiences of coping with climate vulnerabilities.
2. Enabling institutional actors to reflect on how gender norms and power bias and shape climate interventions.
3. Bringing together end-users and stakeholders to collaboratively make sense of any mismatch between ground realities and institutional interventions and exploring the possibilities of bringing together experiential and expert knowledge to inform more inclusive, gender-transformative climate solutions.

A key stereotype we addressed in the research project was the simplistic manner in which gender-climate vulnerabilities are understood. Through early interviews, we identified caste, poverty, and gender as key variables that determine gender-power disparities. Our data show that women are not a homogeneous group, and are not all equally vulnerable to climate impacts. These same variables also shape the structure and culture of institutions and in turn, climate interventions. We noted challenges to participation and representation of marginalized groups and individuals at community levels and in relevant institutions, and how a lack of attention to these issues are reiterated in essentialist narratives on women, creating key barriers to transformative change.

The Significance of a Feminist Political Ecology Lens in Analyzing Intersectional Inequalities and Power Dynamics in Climate-Food Systems Innovations In our research, we adopted a feminist political ecology lens to analyze the historical basis of power and its production and reproduction within institutions. As we note, the combined effects of gender-caste-poverty among respondents is an outcome of a historical and structural inequality rooted in feudal, caste-based control of resources in the research locations. Our data reveal how these values persist, determining why "a son is [still] seen as the family's future", even by women. Patriarchy explains why parents are willing to invest scarce resources in sons rather than in daughters. We found that regardless of caste or class, as well as increasing engagement of women in agriculture, most families believe that daughters belong to the private (household) domain, while sons can and should function in the public sphere. However, not all women (or men) are equal. Landless laborers with little economic or social capital are predominantly from lower castes. They are often excluded from decision-making in the community. These exclusions are further impacted by gender. In situations of increasing male-out migrations, lower-caste, landless women in Bihar are particularly constrained not just by a lack of access to resources, but also by persisting exclusions from information, including climate-adaptation technology and interventions.

The early insights that women are not a homogenous group allowed us to be mindful of intersectional inequalities in the selection of respondents and in the design and use of SenseMaker research frameworks. Our questions not only disaggregate data by caste, gender, poverty, but also probe how these intersections impact both individual experiences, and social interactions. This allowed us to avoid a conventional binary framing of gender inequality, which assumes a universal vulnerability of women, or ignores the experiences of marginalized men across institutions. The research design also required us to look at the historical dimensions of inequality, power hierarchies, institutional structures and cultures, and gender-caste-class blind spots in climate-food systems innovations.

For example, our data show that in rural Gaya, upper caste women do not work in the fields or engage in agricultural tasks. They do not generally come out into the public domain; they engage in domestic tasks and responsibilities. Caste is associated with privilege and status and is a significant factor in determining or restricting the mobility and participation of women outside their homes, especially in agricultural production. Upper caste women's participation in agricultural activities is linked to embarrassment and shame. In upper caste households, outside work (agriculture) is done by others—male and female agricultural laborers. There are no such expectations or restrictions for lower caste women; their poverty, lack of assets and resources requires them to work outside the home, besides managing domestic work. This creates very different types of challenges for women and would also require different types of interventions in relation to climate impacts on agriculture.

This research attempted to answer the following questions: "How do we approach doing science on one of our most important and complex systems – the climate, paying particular attention to how deeply contextual societal norms and biases influence our approach?" and "How do we place gender at the center of new technical innovations in global agricultural research for development (AR4D) and trigger systemic structural gender-transformative change processes across the A4RD institutional landscape?"

Gendered Challenges in Agriculture: Masculinities, Caste, and Class Dimensions The project analyzed the gender-power dimensions of masculinity and femininity at the household, community, and institutional levels. Masculinities that shape institutions and technical interventions are shaped not just by gender but also by caste and class. For instance, despite women's higher involvement in agriculture as laborers due to male-out migration, the notion of masculinity is maintained and reproduced (including the association of machines with men). The connection is maintained so strictly that women farmers wait for male community members to return from different cities to perform the mechanical work if no men from their households are available, or if male labor is expensive.

Structural and Systemic Challenges to Gender-Transformative Change in Climate Interventions The underlying focus of this research was on understanding structural as well as systemic barriers to gender-transformative change. Therefore, we investigated the structure and culture of institutions, as well as probed the combined effects of vulnerabilities by caste, poverty, and gender amongst farmers. Climate interventions oversimplify complexities to single-issue solutions like climate-smart irrigation or weather-based crop insurance, without assessing who may be excluded and why. "What is not counted does not count," leading to technological innovations reduced to simply generating (sex-disaggregated) data. Our research shows that most institutional actors are male, upper-caste Hindus. Most of them felt that the workplace is a neutral space—free from influences of religion, caste, gender, or other biases. Occasionally, some male staff members empathize with women and marginalized groups but point out that the system is not designed to tackle gender equality and social inclusion. More importantly, most staff members are of the view that a focus on those who are "hard to reach" is not always appropriate, efficient, or justifiable. Contrary to narratives, the common understanding here is that climate impacts everyone equally. It particularly impacts agricultural resources, productivity, and ultimately people's livelihood, but at the end of the day, everyone is impacted equally.

Designing and Implementing Contextually-Relevant Climate Solutions Through Plural, Experiential, and Situated Knowledge To transcend interdisciplinary approaches and enable plural, experiential and situated knowledge(s) to inform the design and implementation of contextually-relevant climate solutions, we took several key actions (described below), including assembling an interdisciplinary team.

Equitable partnerships and trust-building with stakeholders and end-users are fundamental principles that guide our use of the SenseMaker tool. We engaged with end-users in the research design by conducting focus group discussions and case study interviews, generating knowledge to feed into the SenseMaker Signification Framework. By partnering with SumArth, we were able to ground the research in a contextually relevant framework and approach and involve local women and men as part of the research team. Our biggest outcome was influencing SumArth on the realities and challenges of intersectional inequalities and vulnerabilities in agriculture.

We followed the feminist research principle of member-checking by taking the data back to the researched communities (Caretta 2016). We are currently analysing the data and preparing to curate meaningful data to create an interactive platform for research dissemination. Our goal is to validate the findings through discussions with end-users and institutional stakeholders separately, then bring both groups together for a townhall discussion to explore more inclusive interventions collaboratively.

By co-designing and implementing the research with the partner organization, SumArth, we have enabled significant learning in a local farmer-producer organization. Our research team, comprising local women and youth drawn from SumArth's membership, were trained and facilitated to pilot the digital ethnographic tool, SenseMaker. SumArth will plan and facilitate knowledge-sharing workshops with relevant groups of stakeholders.

Through qualitative research with key implementation stakeholders, we reflected on gender norms, values, and biases that operate in the workplace. Although our reflections are not yet adequate, they provide some opportunities to move forward on issues of an inclusive workplace.

Designing a Research Tool for Transdisciplinary, Ethnographic Research: Opportunities and Challenges The research tool we used allows for transdisciplinary, ethnographic research through a digital interface. By design, the tool calls for reflexivity at every step of the research. By working closely with the designers of the of the SenseMaker tool, we were able to reflect on deeply contextual challenges in the tool's application—which include intimidation and fear among respondents to have the conversation recorded digitally, as well as more practical challenges relating to how limitations in technology infrastructure and capacity locally delay processes of data collection and analysis.

Reflexivity and introspection were critical elements of our research design, which aimed to ensure the co-design of a research framework that was not a top-down imposition from researchers external to the local context. This required over four months of work to finalize the framework and questionnaires, during which local stakeholders and researchers were engaged and influenced the research design and focus.

Our enumerators were trained in narrative field research techniques, which included framing guiding questions for respondents new to being researched, conducting ways to communicate to gain trust, and being mindful of formal and informal stakeholder groups and gatekeepers with vested interests.

To ensure ethical research, we obtained approval from the IRB at the IWMI, undertook ethical research certification, and trained all enumerators on the need to do no harm. All respondents were briefed on the research objectives and they gave consent before giving the interview. No children or minors were research participants.

Our approach was to be conscious of power dynamics in research projects, and to tackle these challenges by enabling local researchers to lead the research process. However, we also acknowledged our positionality as researchers and the limitations of our backgrounds, which were mostly urban, upper-caste, and literate. During the pilot phase of the project, we took notes and had reflexive sessions each day after the field visit. Interview questions were critically analyzed and reshaped based on feedback from respondents and field experiences.

Despite our efforts, we faced challenges with institutional stakeholders who held authority and power over the research process. We had to work around their availability and agreement to be interviewed, and often, their answers were not reflexive enough, making it difficult to probe further.

Our engagement with our local partner, SumArth, was also not always smooth and tensions relating to insider-outsider, researcher-practitioner issues required significant facilitation by the project team leader. However, through the project, we were able to influence SumArth, which will have a significant impact in the work they lead with other partners, multiple national and international ones. In their own words,

> The project helped us to diversify our team with the emphasis on gender equality. When the project started, we had just two female employees but gradually during the course of our partnership with the project, the number increased to six, and we do see the value of a more gender-balanced team in our organization (a SumArth staff member).

Identified gaps in transformative, participatory approaches and addressing marginalization in research. A key gap that we identified, and are now working towards in the form of a publication, is the lack of know-how on transformative, participatory approaches among researchers and non-researchers. Simply put—how to allow marginalized groups to narrate their stories and experiences and affirmatively engage in problem-solving? Building trust takes a lot of time and effort, starting from making respondents comfortable to sharing deeply personal experiences, particularly negative stories.

Maintaining privacy and individual voices of women interviewees was a significant challenge during the data collection phase. Even while conducting personal interviews, family members or the community would gather around the researcher and respondent. In many cases a male family member remained present throughout the interview. For example, a woman from a backward caste was constantly interrupted by her husband during her interview, who said things like: "She would not know; why are you asking her?" In such situations, interviews had to be paused and resumed in spaces appropriate for the respondent. But we cannot guarantee that this happened in all the cases.

The diversity among sub-caste groups determined that we were not always able to ensure representative voice and engagement of the most marginalized, even though the design of the research was cognizant of the power dynamics of gender, caste, and income. Our sample gender ratio was 60:40 (women to men). We conducted more interviews with women from scheduled castes, scheduled tribes, and extremely backward castes, most of whom were landless and earning a living either as tenant farmers or agricultural laborers.

Cognitive biases of our research team, i.e., subjectivity and positionality in the interpretation of data—is another issue often overlooked by researchers. We have tried to overcome these blind spots by holding several rounds of discussions and ensuring collective wisdom of the group of researchers and local partner participants in analyzing the data and exploring key issues for evidence-based data from an aggregate of individual, personal stories collected from the respondents.

Another key gap we encountered is the lack of data and information on how gender and social exclusion play out in institutional structures and cultures. The institutional actors we interviewed are hardly ever researched and were clearly not used to answering research questions. And yet, so much of what impacts why policies do not deliver rely on what happens within institutions.

10.3.4   Case 4: Time Poverty Among Women Smallholders in Ghana: Implications for Gender Priorities in the Peanut Value Chain

Leland Glenna, Paige Castellanos, Leif Jensen, Janelle Larson, Kaitlin Fischer, Edward Martey, Doris Puozaa, and Richard Oteng-Frimpong

This case study focuses on a gender-integrated FFS conducted for 16 weeks in two communities of Ghana's Northern Region. FFS attendees were participants in a four-year project focused on understanding men and women farmers' time use across seasons. The project's objective was to measure any changes in how men or women spend their time on the farm or in the home after participating in the FFS, with the aim of reducing women's time poverty. Time poverty refers to having insufficient time available to take on new tasks or for rest or leisure due to high agricultural and domestic workloads (Bardasi and Wodon 2010).

The research project titled "Time Poverty Among Women Smallholders in Ghana: Implications for Gender Priorities in the Peanut Value Chain" was funded by the Innovation Lab for Peanut at the University of Georgia through the USAID's Feed the Future Initiative. It was led by a team of researchers at Penn State in the USA and the Savanna Agricultural Research Institute of the Council for Scientific and Industrial Research (CSIR-SARI) in Ghana. Researchers employed quantitative and qualitative methods, including the Abbreviated Women's Empowerment in Agriculture Index (A-WEAI). Survey respondents were asked to respond to questions concerning agricultural production, resources, income, leadership, and time use before and after the FFS. By the end of the project, they provided accounts of their time use during a 24-hour period in the growing, harvest, and dry seasons during six waves of data collection, three before and three after the FFS. Focus group discussions with a sample of participants, men and women, informed FFS implementation. Further qualitative research was conducted with a sample of participating households in the form of in-depth, longitudinal interviews (funded by the Fulbright U.S. Student Program and Penn State's Africana Research Center and African Feminist Initiative).

Peanuts (known as groundnuts in Ghana) are grown by over 90% of agricultural households in northern Ghana (Martey et al. 2015). They are generally considered a "women's crop" (Apusigah 2013), although they are grown by both men and women (Doss 2002; Tyroler 2018). In northern Ghana, women are disproportionately responsible for domestic labor such as cooking, cleaning, and caring for family members (ISG and Ayamga 2017). Three time-use survey waves were used to gain a baseline understanding of men and women's differing responsibilities on the farm and in the home and the time spent on specific activities. Focus group discussions held separately with women and men allowed community members to explain what they view as their greatest time-consuming activities and their greatest constraints on increasing farm production, and to propose solutions to reduce drudgery or other production constraints. The suggestions were then used to design the FFS, which encouraged shifts in culturally entrenched gender roles for the benefit of the entire household. In-depth interviews and three additional time-use surveys were used to understand the extent to which the FFS induced changes in farmers' time use as well as their income, leadership roles, access to resources, and involvement of spouses and other family members in decision-making. Interviews also sought to understand why individual men and women grow the particular crops they do.

The project's FFS consisted of seven technical, farm-based sessions and nine gender-integrated, household-based sessions in each community. The household-based sessions addressed power issues by engaging in open discussions with men and women (separately and together) on topics such as: (1) gender roles and relations, (2) power inequalities and decision-making, (3) crop preferences by gender, (4) skills and ability, (5) conflict and conflict resolution, (6) self-esteem and leadership, (7) time use by men, women, and youth in the household, (8) animal care and responsibilities in the household, and (9) sanitation, hygiene, and nutrition. Each session reflected on individuals' and households' own experiences with a discussion of existing patterns and changing trends within society. Discussions of each topic were grounded in how power differences between men and women have led to and continue to lead to different opportunities and challenges for men and women that are socially constructed, rather than natural.

Participants were encouraged to consider how shifts in certain societal norms could benefit them and their household. For instance, during the session on power inequalities and decision-making, participants discussed the benefits for household members of men and women making decisions together. By the end of the session, one goal was for participants to understand what power is, who has it, and the implications of having power, especially within the household. Discussions centered around the fact that men tend to control access to and use of resources at the household and community levels, and therefore make decisions for the household. Whereas women are usually required to make decisions in consultation with their husbands, men often make decisions without consulting their wives. Participants were introduced to the need for power-sharing and consensus-building in making decisions.

The project took place in two rural communities of northern Ghana selected for their differing proximity to commercial markets and the Northern Region's primary city, Tamale. It was hypothesized that the women and men in each community were likely to engage in varied agricultural and domestic practices due to differing levels of access to markets and labor as well as education. Every household in the two selected communities was included in the project. Households were both monogamous and polygamous and, in cases in which the household head had multiple wives, all wives were included in the research. Understanding the experiences of all wives in polygamous households indicates that even women with the same husband can face very different opportunities and challenges in their everyday lives that affect their time-use, decision-making abilities, and overall well-being.

Every FFS topic was discussed by, and in relation to, both women/femininities and men/masculinities. After establishing a shared understanding of terms and concepts among the group, participants were often divided by gender to consider the topic in greater depth by engaging in a group activity. After the activity, men and women came back together to share what was discussed, ensuring that men and women were active participants, able to share their perspectives and be heard by members of their own gender, and by others. By focusing on improving relationships of all kinds (e.g., between husbands and wives, between co-wives, between women, and between men), the project prioritized change at the individual, household, and community levels.

Every household-based FFS session began with an informal assessment of the prior week's learning goals by asking attendees to answer key questions summarizing the knowledge gained. Participants were also asked to narrate a situation at home or within the community in which they practiced what they had learned. Weekly topics were structured to build upon the topics introduced in earlier weeks. For instance, discussion of the benefits of livestock production for the household incorporated earlier FFS topics on men and women's inequitable access to assets, on women's time poverty, and how these relate to women and men's differing abilities to rear animals and receive the benefits. The final FFS session comprised technical and gender-based learning assessments (and a farmer graduation ceremony).

The post-FFS time use surveys will be combined with in-depth interviews and future participant observation to see if the FFS affected how men and women spend their time. We intend to assess the extent to which men and women have altered gender norms by assisting one another with their work on the farm and in the household.

This project takes an uncommon approach by evaluating the effects of new technology (through the technical FFS sessions) and social innovation (through the gender-based FFS sessions) on men and women smallholder farmers' time use. Many agricultural development projects introduce technologies with no attention paid to their corresponding social innovations, which can result in technologies having limited or even detrimental effects, especially on women (FAO 2023; Theis et al. 2018). This research is attentive to introducing both technical and social innovations and their combined effects. Also unusual is that the research includes the head of the household and all of his wives, contributing to our understanding of the experiences of women in polygamous households in northern Ghana. This is important for understanding the differences that exist between and within households based on the type of marriage within the household and women's social location as the only, first, second, or third wife. Going forward, when introducing these interventions to additional communities, it will be important to recognize that men and women differ not only in whether they are married or not, but in the type of marriage they have.

An interdisciplinary group of researchers both at Penn State and CSIR-SARI spanning the social and natural sciences designed the research project method. This included research instruments such as survey and interview questionnaires as well as the curricula for the FFS and a gender training of research staff. Core team members specialize in rural sociology, agricultural economics, gender, seed science, groundnut technologies, agronomy, and demography. The CSIR-SARI team members and partners led the on-the-ground implementation of the project, in part due to Covid-related travel restrictions imposed on Penn State team members—except for one Penn State rural sociology team member who actively observed the FFS sessions and led in-depth interviews with a sample of project participants. Data analysis and write up are ongoing by quantitative and qualitative researchers on the project team.

The FFS was led by Ghanaian researchers who live and work near the research communities, making them familiar with the general socio-cultural norms in the area. These researchers also have expertise in the spheres in which they contributed to the FFS sessions, namely, good agricultural practices and gender norms in peanut production, processing, and marketing; seed and plant technologies; and gender inequities in Ghanaian households and communities. During the FFS there were open discussions on gender norms and how they affect household members. Participants were shown pictures of how things could be if their perceptions and entrenched beliefs about various subjects changed. The women facilitators of the FFS were themselves examples of what a change in community members' views on gender could lead to, since some of them are from northern Ghana. From its beginning, the project was inclusive by involving participants in the identification and prioritization of their production and time constraints. The team collected baseline quantitative data through surveys and qualitative data through focus groups that informed the design of the remainder of the project, including the FFS intervention.

This research is ongoing, and findings have not yet been shared formally with community members or through peer-reviewed publications. The project plans to share findings from the time use surveys and interviews with community members after the final survey has been conducted and solicit their feedback in a subsequent visit in 2023 (funded by the Society of Woman Geographers and Penn State's Office of International Programs in the College of Agricultural Sciences). Based on observations at the FFS, participant testimonies provided at the FFS graduation, and interviews with a sample of participants, knowledge gained and generated through the FFS has led to behavior change and improved household and community relations for some members of the community.

During the project, the research team embraced learning spaces facilitated by the USAID Innovation Lab network to share preliminary findings with other members of the Innovation Lab, through webinars and presentations. This enabled the research approach and first findings to be shared with individuals new to thinking about gender and with those already incorporating gender into their research, such as other USAID Feed the Future Initiative grantees involved in the Affinity Group for Gender within the USAID Innovation Lab Cross-cutting Theme Community of Practice. The project is contributing to efforts to motivate institutional change across the Innovation Lab network by illustrating the value of research rooted in social sciences (Marter-Kenyon 2022). Most other USAID agricultural research and development efforts are rooted in natural sciences, with no social science component beyond economics. Additionally, findings have been shared in seminars with other CSIR-SARI researchers and with the Ghana Groundnut Working Group, supported by USAID's Peanut Innovation Lab and comprised of researchers, farmers, aggregators, and processors in Ghana's peanut sector. The team is also fostering partnerships with organizations and practitioners in northern Ghana with the intention of scaling up the gender-integrated FFS method by bringing it to additional communities across the region.

The research team held a Gender and Agriculture Workshop for SARI employees involved in the project, such as enumerators collecting survey data, and employees involved in other socio-economics projects. The workshop focused on why gender matters to agricultural research, how to conduct feminist research, and incorporating gender into monitoring and evaluation and outreach. Survey findings have been shared at various seminars, workshops, and conferences with stakeholders from institutions working in the agricultural research and development arena. The present research findings have shown, for the first time in Ghana, that inequalities do exist within households based on household members' gender identity and type of marriage (monogamous or polygamous), and that for women, being an only, first, second, or third wife matters.

Reflecting on the power dynamics within the research relationship was necessary throughout the project. The team applied participatory methods and encouraged experiential knowledge sharing among the core research team and partners during the design and implementation of the project. For instance, before the FFS, community members were given the opportunity to discuss their production constraints, prioritize them, and suggest possible solutions. These suggestions set the stage for curriculum development and technology introduction. The research team recognized that its ability to provide research participants with small benefits, such as snacks during the FFS sessions, could generate tension within the community, so what snacks were provided, when, and in what quantity was given careful consideration; snacks were handed over to the group leaders for distribution to participants. The researchers sought to understand participants' individual experiences and promote principles that would improve participants' well-being, rather than prescribe moral boundaries, which as non-community members, would have been inappropriate. The research team ensured the use of simple language and played facilitator instead of teacher roles during the FFS sessions.

The research used both reflexive and introspective approaches. The team employed methods that allowed room for modifications as long as they did not alter the ultimate goals of the project. Bearing in mind the differences in backgrounds of the research team and participants, issues were extensively discussed and the best courses of action taken. Efforts were made to respect the views of all partners and the same respect was insisted on among farmers during the FFS. The research team plans to continue conducting research in northern Ghana. In the future, they will increase communication among Ghanaian and American team members—which was limited by the barriers to travel presented by Covid—to improve the project's reflexivity. The team could have gone even further to interrogate the power dynamics within the research relationship. Another project limitation is that this research relies on a binary notion of gender that only recognizes two genders: man and woman. It fails to account for gender as a spectrum.

10.4   Moving Beyond Gender-Transformative Approaches

The authors started out this chapter by suggesting some reasons why the use of feminist research approaches in agriculture is not very common. These included epistemological and methodological differences between and within organizations, low staff capacities, a strong focus on shorter-term outcomes and a propensity to overuse instrumental approaches, and an overall resistance within agricultural research organizations to embrace such approaches. We also presented some details on what constitutes feminist research in agriculture, and using the conceptual framework developed by the IWDA (2017), we highlighted the key or mandatory components of doing rigorous feminist research. The framework provided a valuable guide to structure and review the four case studies that were included in this chapter. The guiding questions in Appendix that were used by authors to write up the different case studies can also be used to guide the design, implementation, and monitoring and evaluation of future feminist research in agriculture.

The case studies illustrate that each research project focused their attention on understanding gendered power relations and discriminatory institutions, especially informal norms. The projects mainly adopted non-traditional research approaches that aimed to empower research participants and create a safer space for critical reflection on and action to address the root causes of gender inequalities. Each research project studied masculinities and femininities within agriculture and how these interacted with other contextually relevant social categories to benefit or disadvantage certain groups of people. All projects followed a transparent process to ensure ethical engagement with research participants and reported that research team members played primary roles as facilitators rather than as "objective" observers or science experts. This seemingly shifted the power dynamics between researchers and research participants to help ensure that the latter's voices were heard and informed project design, implementation, and monitoring and evaluation.

Each case used several other feminist research principles when carrying out their projects. Using an intersectional lens, most of the cases reported that they captured the diversity of women and men, going well beyond viewing women and men as homogeneous groups. Each project clearly embodied interdisciplinarity. Using both qualitative and quantitative methods and perspectives from a multitude of disciplines, the different research projects generated a rich body of evidence to enable a range of stakeholders to reflect on and come up with ways to address gender inequalities and start gender-transformative change. In terms of accountability, the research projects created alliances and learning spaces with various actors, which ranged from donor agencies to community-based organizations. While not all projects reported that they completed implementation and disseminated their results, gender-transformative change processes were set up and are bearing fruit across the four cases. Using feminist research approaches within each project, these cases highlight how a range of agricultural research work can be implemented so as to help bring about women's empowerment and gender-equal development outcomes.

Our review of the four case studies suggests that the use of feminist research principles is likely more evident in agricultural research projects that adopt gender-transformative approaches than in those that are gender-responsive or that mainstream gender in their project design and implementation plans to ensure they meet gender targets and the practical needs of women and men.[4] While the similarities between gender-transformative and feminist research approaches make conceptual sense, Mullinax et al. (2018) point out that these approaches are different, and that gender-transformative research may or may not be feminist. Both approaches include power as their central focus, but feminist research challenges the use of mainstream research approaches by bringing in participants as co-researchers (or experts), rebalancing power relationships between the researcher and participants, and rejecting the notion that research is value-free and objective.

Moving beyond the use of gender-transformative approaches in agricultural research is a necessary next step to alter gendered power relations and discriminatory institutions, and to fundamentally change how agricultural research is carried out. To create an enabling environment for the use of feminist research approaches in agriculture, the larger agricultural research system clearly needs to change. Mullinax et al. (2018) maintain that broader institutional systems can support such change by aligning their internal organizational systems and policies with feminist

---

[4] See the following link for information on what gender mainstreaming entails https://eige.europa.eu/gender-mainstreaming/what-is-gender-mainstreaming

principles, adopting frameworks that integrate gender-transformative thinking in all aspects of organizations, and building staff capacities to do feminist research. They also advocate for donors to fund more agricultural research that embodies feminist principles, including research that reshapes how knowledge gets created and who owns the research, research that is more action oriented and leads to significant change at different levels, and building the capacity of researchers to implement that which is required for change to happen. The issue of sustained funding by donors to support feminist research in agriculture is also important as the first case study showed that, while financial support to carry out the research was provided for five years, it was not sustained into the longer term because donors prioritized funding for other purposes. Such longer-term commitments on the part of donors, but also researchers and their organizations, are needed to ensure that feminist research can help facilitate deeper level, transformative change.

As more organizations and donors support these efforts, the use of feminist research approaches in agriculture may increase and help shift how we do agricultural research and prioritize women's empowerment and gender equality as goals in and of themselves rather than mechanisms for increased productivity, food security, and the like. Nonetheless, generating an evidence base to showcase how the use of feminist research approaches in agriculture can bring about positive development outcomes is important as one means to gain institutional support at various levels. Arguably, this evidence base is relatively low given the lack of use of feminist research approaches in agricultural research, and yet this evidence base can only increase once institutional support creates an enabling environment for researchers across agricultural research topics (e.g., crop and animal breeding, agronomy, natural resource management, value chain development, consumer food preferences) to use feminist research principles when designing and implementing their research.

10.5   Conclusion

This chapter highlights the importance of using feminist research approaches within agriculture as a promising means to achieve equitable and sustainable transformations of agrifood systems. In this chapter we put forth several reasons to help explain the current lack of use of feminist research approaches in agriculture, while acknowledging the dearth of human resources in organizations with the capacities to do feminist research. We described what constitutes feminist research in agriculture by highlighting several frameworks and principles that researchers use to inform the design and implementation of their feminist research. We also provided detailed case study examples that carried out rigorous feminist research to showcase how feminist research can be implemented in an agricultural context. In practice, it is imperative to recognize that using feminist research approaches requires realistic timeframes with budgetary commitments as prerequisites for achieving gender equality as opposed to the usual focus on short-term outcomes. The latter is associated with reliance on instrumental approaches that insufficiently address the underlying causes of gender inequalities.

We end this chapter with an appreciation that resistance to using feminist research approaches within and across organizations is and will continue to be common. Interrogating power relations that disadvantage certain groups of women and marginalized groups in agriculture and helping facilitate ways to change discriminatory norms that perpetuate or justify their subordination, is not an easy task to carry out within the agriculture sector. Across the globe, agriculture is male-dominated in organizations, especially at management or leadership levels. Including women and men farmers and other stakeholders as co-researchers to inform the design, implementation, and monitoring and evaluation of the research, goes against traditional agricultural research that usually assumes scientists are the ones who come up with solutions for farmers and other value chain actors. Mainstreaming the use of unconventional, or radical approaches, including feminist ones, in agricultural research starts at the top of organizations. To transform agrifood systems to be more inclusive, gender-equal, and socially sustainable, leaders need to institutionalize feminist research approaches across their organizations. Without commitment from the top, change will be limited and slow.

Acknowledgements Funding to support the writing of this chapter by the lead author was provided by the Harnessing Gender and Social Equality for Resilience in Agrifood Systems (HER+) Initiative. Thanks to all funders who supported the HER+ Initiative through their contributions to the CGIAR Trust Fund https://www.cgiar.org/funders/. The Case 2 research team, consisting of Janelle Larson, Paige Castellanos, Leif Jensen, Carolyn Sachs, Arie Sanders, Alfredo Reyes, and Hazel Velasco would like to thank Víctor Vargas, Sheyla Ramos, and the Horticulture Innovation Lab at Zamorano for their valuable contributions to this research. In particular, we are thankful to AMIR's staff, especially doña Pascualita García, Juliana Meza, Olga Pérez, Mercedes García, and all the participants who shared their time with us. This study was supported by the generous support of the American people through the Horticulture Innovation Lab with funding from the U.S. Agency for International Development, as part of the U.S. government's global hunger and food security initiative called Feed the Future under Prime Award No. 09-002945-105 and Sub Award No. 5257-ZU-UCD-5105. The content of Case 2 is the responsibility of the authors and does not necessarily reflect the views of USAID, the United States Government, or any donor. The successful execution of the research presented in Case 3 was made possible thanks to the generous funding provided by the CGIAR GENDER Impact Platform. We wish to express our utmost appreciation to our collaborative partner Anna Panagiotou from Cynefin Co. Additionally, we would like to extend our gratitude to Upandha Udalagama, an IWMI consultant, for her invaluable research assistance. The Case 4 research team, consisting of Leland Glenna, Paige Castellanos, Leif Jensen, Janelle Larson, and Kaitlin Fischer at Penn State and Edward Martey, Doris Puozaa, and Richard Oteng-Frimpong at CSIR-SARI, thanks the farmers and their community leaders who generously gave their time to take part in the project. They also thank the CSIR-SARI enumerators who collected the survey data and all those who were instrumental in implementing the farmer field school including Bernice Wadei, Eleazar Ofosu Krofa, Salamatu Mahamadu, Akemo-M. Rasheed Ignatius, Abdulai Abdul Malik, Alidu Kubura, Constant Obranie, Anarfo Talata Brandina, and Alhassan Chinto. This project was made possible by the generous support of the American people through USAID's Cooperative Agreement No. 7200AA 18CA00003 to the University of Georgia for U.S. Feed the Future Innovation Lab for Peanut. The contents included in Case 4 are the responsibility of the authors and do not necessarily reflect the views of USAID or the United States Government.

Appendix: Guiding Questions Used to Develop the Case Studies

Element 1: Building feminist knowledge of women's lives

- How did the research investigate or address the harmful and positive impacts of gender-based stereotypes, and/or understand differences between women and men and why these differences exist?
- How did the research address power issues, and how power operates and affects individuals and larger groups of people in communities, and the fact that gendered power relations are grounded in historical contexts?
- How did the research examine the experiences of women in their diversity, and the impact of intersectional identities on women's lives? How did the research reject simple binaries? How did the research try to hear multiple voices throughout the research process and/or give voice to women within different social groups and/or contexts?
- Did the research study both women/femininities and men/masculinities and their construction and interaction? Please explain how.
- Was the research based on a theory of change that focused on assessing incremental progress toward gender transformative change? Please explain.

Element 2: Accountable for how research is conducted

- How did the research pursue new and neglected research questions?
- How did the research embody interdisciplinarity in its design/implementation/analysis/write up?
- How was the research grounded in a commitment to do no harm?
- How was the research methodologically rigorous in its use of a range of feminist participatory research methods? How did the research try to understand what women want through the research process? How was the research intentional in its design to ensure it was always leading to action?
- How were the findings disseminated and did the dissemination strategy aim to elevate marginalized voices and connect participants into important spaces of influence? How did the research empower participants to use findings to create change in their communities?
- Did the research team create alliances and learning spaces to increase research utilization?
- How did the research try to convince audiences of the realities of gender inequality and communicate how gender transformative change can happen? How was the research used to strengthen advocacy to transform and influence policy?
- Was the research team reflexive and introspective during the life of the research/research project? Did the team appreciate that research is neither value-free nor disinterested? Did the research consider and value different ways of knowing?

Element 3: Committed to ethical collaboration

- How did the research follow a transparent process to ensure ethical engagement with research partners?
- How did the research team interrogate the multiple power dynamics of the research relationship?

Element 4: Seeking for transformative impact on the causes of gender inequality

- At what levels or with which groups did the research focus to have a transformative impact on the causes of gender inequality (e.g., individual and/or organizational level, with the feminist movement, at societal level, and/or in the ways knowledge production is carried out and research methods are developed/used)? Please explain how transformative change occurred and was documented throughout the course of the research.

References

Allen KR (2022) Feminist theory, method, and praxis: toward a critical consciousness for family and close relationship scholars. J Soc Pers Relat 40(3):1–38. https://doi.org/10.1177/02654075211065779

Apusigah AA (2013) Women's agency and rural development in Northern Ghana. In: Yaro JA (ed) Rural development in northern Ghana. Nova Science Publishers, New York, pp 147–168

Bardasi E, Wodon Q (2010) Working long hours and having no choice: time poverty in Guinea. Fem Econ 16(3):45–78. https://doi.org/10.1080/13545701.2010.508574

Cadesky J (2020) Built on shaky ground: reflections on Canada's feminist international assistance policy. Int J Canada's J Glob Policy Anal 75(3):298–312. https://doi.org/10.1177/0020702020953424

Caretta MA (2016) Member checking: a feminist participatory analysis of the use of preliminary results pamphlets in cross-cultural, cross-language research. Qual Res 16(3):305–318. https://doi.org/10.1177/1468794115606495

Choudhury A, Castellanos P (2020) Empowering women through farmer field schools. In: Sachs C, Jensen L, Castellanos P, Sexsmith K (eds) Routledge handbook of gender and agriculture. Routledge, New York, pp 251–262. https://doi.org/10.4324/9780429199752-23

Cole SM, Kantor P, Sarapura S, Rajaratnam S (2014a) Gender transformative approaches to address inequalities in food, nutrition, and economic outcomes in aquatic agricultural systems in low-income countries. CRP AAS, Penang. Program Working Paper: AAS-2014-42. https://hdl.handle.net/10568/68525

Cole SM, van Koppen B, Puskur R, Estrada N, DeClerck F, Baidu-Forson JJ, Remans R, Mapedza E, Longley C, Muyaule C, Zulu F (2014b) Collaborative effort to operationalize the gender transformative approach in the Barotse Floodplain. CRP AAS, Penang. Program Brief: AAS-2014-38. https://hdl.handle.net/20.500.12348/222

Cole SM, Puskur R, Rajaratnam S, Zulu F (2015) Exploring the intricate relationship between poverty, gender inequality, and rural masculinity: a case study from an aquatic agricultural system in Zambia. Cult Soc Masculinities 7(2):154–170. https://hdl.handle.net/20.500.12348/235

Cole SM, McDougall C, Kaminski AM, Kefi AS, Chilala A, Chisule G (2018) Post-harvest fish losses and unequal gender relations: drivers of the social-ecological trap in the Barotse Floodplain fishery. Zambia Ecol Soc 23(2):18. https://doi.org/10.5751/ES-09950-230218

Cole SM, Kaminski AM, McDougall C, Kefi AS, Marinda P, Maliko M, Mtonga J (2020) Gender accommodative versus transformative approaches: a comparative assessment within a post-harvest fish loss reduction intervention. Gend Technol Dev 24(1):48–65. https://doi.org/10.108 0/09718524.2020.1729480

Cole SM, Barker-Perez E, Rajaratnam S, McDougall C, Kato-Wallace J, Muyaule C, Kaunda P, Puskur R, Longley C, Mulanda A, Mulanda J, Patel AE (2021) Changes in intra-household decision-making powers: effects of a gender transformative approach in saving groups in the Barotse Floodplain, Zambia. CGIAR Research Program on Fish Agri-Food Systems, Penang. https://hdl.handle.net/20.500.12348/5103

Davis K, Nkonya E, Kato E, Mekonnen DA, Odendo M, Miiro R, Nkuba J (2012) Impact of farmer field schools on agricultural productivity and poverty in East Africa. World Dev 40(2):402–413. https://doi.org/10.1016/j.worlddev.2011.05.019

Dierksmeier B, Cole SM, Teoh SJ (2015) Focal community profiles for Barotse hub, Zambia. CGIAR Research Program on Aquatic Agricultural Systems, Penang. Program Report: AAS-2015-06. https://hdl.handle.net/20.500.12348/481

Doss CR (2002) Men's crops? Women's crops? The gender patterns of cropping in Ghana. World Dev 30(11):1987–2000. https://doi.org/10.1016/S0305-750X(02)00109-2

EIGE (European Institute for Gender Equality) (2016) Institutional transformation. Gender Mainstreaming Toolkit. European Union, Luxembourg. https://eige.europa.eu/publications/institutional-transformation-gender-mainstreaming-toolkit

Elias M, Cole SM, Quisumbing A, Meinzen-Dick R, Perez A-M, Twyman J (2021) Assessing women's empowerment in agriculture. In: Pyburn R, van Eerdewijk A (eds) Advancing gender equality through agricultural and environmental research: past, present and future. IFPRI, Washington, DC. https://doi.org/10.2499/9780896293915

FAO (2022) Joint Programme on gender transformative approaches for food security and nutrition. Food and Agriculture Organization, Rome. https://www.fao.org/joint-programme-gender-transformative-approaches/en

FAO (2023) The status of women in agrifood systems. Food and Agriculture Organization, Rome. https://doi.org/10.4060/cc5343en

Farhall K, Rickards L (2021) The "gender agenda" in agriculture for development and its (lack of) alignment with feminist scholarship. Front Sustain Food Syst 5:1–15. https://doi.org/10.3389/fsufs.2021.573424

Feldman S (2018) Feminist science and epistemologies: key issues central to GENNOVATE's research program. GENNOVATE resources for scientists and research teams. CDMX, CIMMYT, Mexico. https://hdl.handle.net/10568/106762

Godtland EM, Sadoulet E, de Janvry A, Murgai R, Ortiz O (2004) The impact of farmer field schools on knowledge and productivity: a study of potato farmers in the Peruvian Andes. Econ Dev Cult Chang 53(1):63–92. https://doi.org/10.1086/423253

Ikävalko E, Kantola J (2017) Feminist resistance and resistance to feminism in gender equality planning in Finland. Eur J Women's Stud 24(3):233–248. https://doi.org/10.1177/1350506817693868

ISG (International Solutions Group), Ayamga M (2017) Ghana country report: Ghana POWER baseline study. ActionAid International and ActionAid Ghana. https://aidstream.org/files/documents/POWER-Ghana-Baseline-Country-Report-20171201031250.pdf

IWDA (International Women's Development Agency) (2017) Feminist research framework. https://iwda.org.au/resource/feminist-research-framework/

Jenkins K, Narayanaswamy L, Sweetman C (2019) Introduction: feminist values in research. Gend Dev 27(3):415–425. https://doi.org/10.1080/13552074.2019.1682311

Kabeer N (2007) Marriage, motherhood and masculinity in the global economy: reconfigurations of personal and economic life. IDS Working Paper 290

Kabeer N (2016) Gender equality, economic growth, and women's agency: the "endless variety" and "monotonous similarity" of patriarchal constraints. Fem Econ 22(1):295–321. https://doi.org/10.1080/13545701.2015.1090009

Kaminski AM, Cole SM (2018) Building a case for using participatory and gender-aware approaches in post-harvest fish loss assessments and value chain interventions. Proceedings from the Meeting of Experts in Support of Fish Safety, Technology and Marketing in Africa – Elmina, Ghana, 14 to 16 November 2017. Food and Agriculture Organization, Rome. http://www.fao.org/3/CA0374B/ca0374b.pdf

Kantor P (2013) The contribution of gender transformative approaches to value chain research for development. Livestock and Fish Brief 2. Nairobi, ILRI. https://cgspace.cgiar.org/handle/10568/33795

Kantor P, Apgar M (2013) Transformative change in the CGIAR research program on aquatic agricultural systems. Penang, Malaysia. Program Brief: AAS-2013-25. https://digitalarchive.worldfishcenter.org/handle/20.500.12348/794

Kato-Wallace J, Cole SM, Puskur R (2016) Coalitions to achieve gender equality at scale: gender development and coordinating subcommittees and networks as drivers of change in Zambia. Promundo-US/WorldFish, Lusaka/Washington, DC. https://promundoglobal.org/resources/coalitions-to-achieve-gender-equality-at-scale/

Kiguwa P (2019) Feminist approaches: an exploration of women's gendered experiences. In: Laher S, Fynn A, Kramer S (eds) Transforming research methods in the social sciences: case studies from South Africa. Wits University Press, pp 220–235. https://www.jstor.org/stable/10.18772/22019032750.19

Kingston AK (2020) Feminist research ethics. In: Iphofen R (ed) Handbook of research ethics and scientific integrity. Springer, Cham, pp 531–549. https://doi.org/10.1007/978-3-030-16759-2_64

Larson JB, Castellanos P, Jensen L (2019) Gender, household food security, and dietary diversity in western Honduras. Glob Food Sec 20:170–179. https://doi.org/10.1016/j.gfs.2019.01.005

Leung L, Miedema S, Warner X, Homan S, Fulu E (2019) Making feminism count: integrating feminist research principles in large-scale quantitative research on violence against women and girls. Gend Dev 27(3):427–447. https://doi.org/10.1080/13552074.2019.1668142

Manning J (2018) Becoming a decolonial feminist ethnographer: addressing the complexities of positionality and representation. Manag Learn 49(3):311–326. https://doi.org/10.1177/1350507617745275

Marter-Kenyon J (2022) How do we close the gender gap? (hint: ask a social scientist). AgriLinks Blog https://agrilinks.org/post/how-do-we-close-gender-gap-hint-ask-social-scientist

Martey E, Wiredu AN, Oteng-Frimpong R (2015) Baseline study of groundnut in Northern Ghana. LAP Lambert Academic Publishing, Saarbrücken

McDougall C, Cole SM, Rajaratnam S, Brown J, Choudhury A, Kato-Wallace J, Manlosa A, Meng K, Muyaule C, Schwartz A, Teioli H (2015) Implementing a gender transformative research approach: early lessons. In: Douthwaite B, Apgar JM, Schwarz A, McDougall C, Attwood S, Senaratna Sellamuttu S, Clayton T (eds) Research in development: learning from the CGIAR research program on aquatic agricultural systems. CGIAR Research Program on Aquatic Agricultural Systems, Penang. Working Paper: AAS-2015-16. https://hdl.handle.net/10568/75552

McDougall C, Newton J, Kruijssen F, Reggers A (2021) Gender integration and intersectionality in food systems research for development: a guidance note. CGIAR, Penang. Research Program on Fish Agri-Food Systems. Manual: FISH-2021-26. https://www.sei.org/publications/gender-integration-food-systems/

Mullinax M, Hart J, Garcia, AV (2018) Using research for gender-transformative change: principles and practice. International Development Research Center (IDRC) and American Jewish World Service (AJWS) https://ajws.org/wp-content/uploads/2019/05/Gender-Transformative-Research.pdf

Njuki J, Melesse M, Sinha C, Seward R, Renaud M, Sutton S, Nijhawan T, Clancy K, Thioune R, Charron D (2022) Meeting the challenge of gender inequality through gender transformative research: lessons from research in Africa, Asia, and Latin America. Can J Dev Stud/Revue canadienne d'études du développement. https://doi.org/10.1080/02255189.2022.2099356

Ozkazanc-Pan B (2012) Postcolonial feminist research: challenges and complexities. Equal Divers Incl 31(5–6):573–591. https://doi.org/10.1108/02610151211235532

Park CMY, Picchioni F, Franchi V (2021) Feminist approaches to transforming food systems: a roadmap towards a socially just transition. Agric Dev 42. https://www.researchgate.net/publication/351599945_Feminist_approaches_to_transforming_food_systems_a_roadmap_towards_a_socially_just_transition

Podems DR (2010) Feminist evaluation and gender approaches: there's a difference. J Multidiscip Eval 6(14):1–17. https://journals.sfu.ca/jmde/index.php/jmde_1/article/view/199

Rajaratnam S, Cole SM, Fox KM, Dierksmeier B, Puskur R, Zulu F, Teoh SJ, Situmo J (2015) Social and gender analysis report: Barotse Floodplain, Western Province, Zambia. CGIAR Research Program on Aquatic Agricultural Systems, Penang. Program Report: AAS-2015-18. https://hdl.handle.net/20.500.12348/241

Rajaratnam S, Cole SM, Kruijssen F, Sarapura S, Longley C (2016) Gender inequalities in access to and benefits derived from the natural fishery in the Barotse Floodplain, Zambia, Southern Africa. Asian Fish Sci J, Special Issue 29S:47–69. http://www.asianfisheriessociety.org/publication/abstract.php?id=1110

Rao N (2005) Questioning women's solidarity: the case of land rights, Santal Parganas, Jharkhand, India. J Dev Stud 41(3):353–375. https://doi.org/10.1080/0022038042000313282

Sanders A (2021) Ensamblando la cadena de cultivos de alto valor en el occidente de Honduras. Eutopía: Revista de Desarrollo Económico Territorial 20:97–112. https://doi.org/10.17141/eutopia.20.2021.5160

Shameem N (2021) Rights at risk: time for action. Observatory on the universality of rights trends report. Association for Women's Rights in Development (AWID), Toronto. https://www.awid.org/ours-2021

Theis S, Lefore N, Meinzen-Dick R, Bryan E (2018) What happens after technology adoption? Gendered aspects of small-scale irrigation technologies in Ethiopia, Ghana, and Tanzania. Agric Hum Values 35:671–684

Tickamyer AR (2020) Feminist methods and methodology in agricultural research. In: Sachs C, Jensen L, Castellanos P, Sexsmith K (eds) Routledge handbook of gender and agriculture. Routledge, New York, pp 239–250

Travis C, Garner E, Pinto Y, Kayobyo G (2021) Gender capacity development in agriculture: insights from the GREAT monitoring, learning, and evaluation system. J Gend Agric Food Secur   6(2):19–40.   https://agrigender.net/views/GREAT-monitoring-learning-evaluation-JGAFS-622021-2.php

Tyroler C (2018) Gender considerations for researchers working in groundnuts. USAID. https://ftfpeanutlab.caes.uga.edu/content/dam/caes-subsite/ftf-peanut-lab/documents/peanut-lab/Gender%20Considerations.pdf

Webster S, Lewis J, Brown A (2014) Ethical considerations in qualitative research. In: Ritchie J, Lewis J, McNaughton Nicolls C, Ormston R (eds) Qualitative research practice: guide for social science students and researchers. SAGE Publications Ltd, Los Angeles, pp 77–107

Wong F, Vos A, Pyburn R, Newton J (2019) Implementing gender transformative approaches in agriculture. A discussion paper for the European Commission. CGIAR Collaborative Platform for Gender Research, Amsterdam. https://www.kit.nl/wp-content/uploads/2020/08/Gender-Transformative-Approaches-in-Agriculture_Platform-Discussion-Paper-final-1.pdf

Chapter 11
Critical Reflections Towards Re-politicizing Gender in Agriculture

Jemimah Njuki, Hale Ann Tufan, Vivian Polar, and Hugo Campos

Abstract Advancing gender equality in agriculture is an inherently political process. In providing a conclusion to an excellent set of chapters that provide critical reflections on how to navigate this political process, we call for a paradigm shift-from fixing women to fixing systems that are inequitable, unjust and undemocratic. We propose four critical steps for doing this. First is to acknowldege the gendered hierarchies and power dynamics built into the agriculture sector. Second, recognize the interconnectedness of women's lives. Third, bring women's rights to the fore. And fourth, re-engineer patriarchal organizations and systems to address gender-based discrimination.

As Editors of this volume, we thought long and hard about what we wanted to have in this last chapter of the book. We knew that we did not want to have a typical conclusion, rehashing some of what the excellent authors of this volume have so eloquently said. What we wanted to do is summarize a set of key takeaways we would like to highlight, that emerged as we put the book together.

We have long known, as many others in the sector have too, that advances towards gender equality in agricultural development requires multiple complex challenges to be addressed simultaneously. But what we often neglect, is that the first step is to acknowledge that this is inherently a **political process** strongly dependent on a multiplicity of power dynamics. We call on the agricultural development community to emerge from the sleepy confines of denial. We have a problem. Agricultural development business as usual is not working for women. Most

J. Njuki (✉)
United Nations Entity for Gender Equality and the Empowerment of Women and Girls, UN Women, New York, NY, USA
e-mail: jemimah.njuki@unwomen.org

H. A. Tufan
Global Development, Cornell University, Ithaca, NY, USA

V. Polar · H. Campos
International Potato Center, Lima, Peru

© The Author(s) 2025
J. Njuki et al. (eds.), *Gender, Power and Politics in Agriculture*,
https://doi.org/10.1007/978-3-031-60986-2_11

interventions barely scratch the surface, often treating the symptoms in what can be called a "painkiller syndrome", while failing to honestly acknowledge and address the deep structural inequalities in agricultural and food systems.

This book on re-politicizing gender in agriculture analyzes how gender has continued to be instrumentalized in the agricultural research for development discourse over the years. Gender has become another buzzword, something transversal within agriculture that is diluted and ticked as a box. More often than not, it has become part of technical innovations, approaches and processes. Women on the other hand continue to be seen as receivers and users of technologies and not critical leaders and thinkers in transforming agriculture and food systems. We are still talking about "half of the world's agricultural technology users are women" rather than "why are so few decisions still made for/with or by women". We continue to try and "fix women", rather than "fixing the systems" that are inequitable, unjust and undemocratic.

In **accepting the position that striving for gender equity is an inherently political process** we must also accept that deep-rooted norms, stereotypes and even institutional frameworks constrain women's rights and shape their everyday lives. We must return to rights-based arguments, rather than those rooted in efficiency and productivity. We must acknowledge that the legacy of colonialism and patriarchy in its varied manifestations are very much alive and as effectively discriminatory as ever. This is reflected in lack of political will from governments to revise and abolish discriminatory laws or legal frameworks, develop gender transformative policies and designate funding to advance gender equality. For example, available evidence shows that 67 countries around the world lack laws that prohibit direct and indirect discrimination against women and in 53 countries, the law does not mandate equal remuneration for work of equal value (ICF International 2023). The outcome of this lack of political will cascades into a diversity of areas in the agricultural sector. As Tufan et al., conclude in the chapter that explores different narratives on land and agricultural development, we need to plant the seeds of change by starting from dialogue of critical and liberal approaches to address core issues such as land relations that are a basic bottleneck for equitable development through agriculture. Critical theory, intersectional and decolonial approaches must be brought to the front in the design of agricultural development interventions not only to achieve final development goals but as tools to plant critical thinking in the minds and practices of individuals and institutions. It is through iterative questioning of current development models that we can foster creative destruction of current paradigms and the gestation of new patterns to address development from the perspective of the multiple "other actors" in society.

Understanding current production paradigms and history of development thinking is basic to catalyze change towards gender equity in agricultural development. As Wong highlights, hegemonic masculinity is reproduced over time in the agricultural sector through the ways of working and gradually through the institutional practices and frameworks. If the structures that birth the majority of agricultural innovation are rooted in patriarchal paradigms, it is not at all surprising that mostly men participate and lead research and development in the agricultural research for development sector. As Njuki et al., highlight in their chapter on navigating patriarchal politics in agriculture, institutions have established frameworks that systematically limit the extent and quality of women's engagement. This partially explains why there are less women today involved in designing, leading, adopting and benefiting from agricultural research outputs and development practice. A clear example is presented by Kansanga and Dinko who use mechanization as a case to exemplify how technology design reinforces existing inequalities and policy frameworks, thus enhancing the inequality gap.

Embracing Diversity and Complexity of Social Structures Is Essential Conscious of the need for inclusive and actor led processes, participatory methods emerged as an alternative to shift the agricultural research and development trajectory towards more community driven innovation and processes. However, as Ashby mentions in her chapter on participatory research, AR4D bureaucracies with a supply driven focus and an efficiency led paradigm, have limited the full expression of participatory research. In an attempt to facilitate development, there has been a historic trend towards simplification and dichotomy—men and women, large scale and small scale, rich and poor. We acknowledge the challenge of managing a complex reality but dichotomizing and simplification can obscure important segments of the population. As Tavenner et al., mention, the lack of attention to diversity and intersectionality can be the reason why large portions of AR4D interventions are not able to bring lasting, sustainable change in the desired direction. Other approaches to simplification such as the development of tools to standardize social analysis and foster replicability can run the risk of losing the critical thinking component and transformative nature. As Cullen et al., highlight, it is more important for social sciences to play a critical and reflexive role which entails questioning objectives, methods and outcomes of the research process, instead of merely serving technical agendas of the biological sciences in agriculture.

Moving Beyond the Use of Gender-Transformative Approaches in Agricultural Research Is a Necessary Next Step As Cole et al. highlight, the use of gender transformative approaches and feminist research principles has generated positive outcomes, we need to move a step forward to address gendered power relations and discriminatory institutions to fundamentally change how agricultural research is conducted and how development is conceived and operationalized.

Empowerment is not something to be bestowed or conferred, it is a process of inner change that implies navigating multiple expressions of power dynamics. While considerable attention has been devoted to addressing and measuring women's empowerment in the last decade, we must acknowledge the fact that, as Polar and Poole mention, measuring itself holds an intrinsic bias and serves a specific purpose. We must enquire about the nature of the underlying measurement agenda and ask who, how and why measurement is addressed; what are we missing in the measurement scenario and how can results from measurement help challenge existing power relations and conservative social and cultural norms. Power is at the core of social processes and as such needs to be part of the social analysis and needs to be openly addressed.

To catalyze change, interventions need to shift current development paradigms. Delving into the nuances of history, masculinities and hegemonic practices can help us identify the basic bottlenecks, how they evolved over time, as well as trends and opportunities for change that can be incorporated in the design of interventions that shift the paradigm.

While addressing all these issues is complex, there are a few critical steps that can be taken to move us in the right direction:

1. **Acknowledge the gendered hierarchies and power dynamics that are built into agriculture** and broader food systems scenario, including financing, research, extension systems, analyze them and work towards changing them as a collective of actors working in agricultural systems. While the rest of the world is moving towards thinking about alternative economic models, there has been little discussion within the agriculture sector, of what alternative models in the sector that work for women and diverse groups are, and how to move towards the systematic implementation of these models. It is especially important for the gender and agriculture community to rethink the instrumental concept of "gender equality as smart economics" which has endorsed women's participation as good for the economy, valuing women for their useful contribution to alleviating hunger, and to contributing to household production and nutrition and how this approach has led to the neglect of our understanding of how women continue to operate in patriarchal systems that have not evolved and that do not serve them.

2. **Recognize the interconnectedness of women's lives.** Within the agriculture sector, there needs to be more focus on the interconnectedness of women's lives for example by linking the work we do on gender research to poverty dynamics. By 2030, it is projected that 340 million women will live in poverty, currently 1 in 3 women face gender-based violence, and women are carrying a disproportionate share of unpaid care work. Polices and programs in the agriculture sector need to connect more closely with care policies, macro-economic policies, social protection policies with a rights-based focus to ensure better lives for women and girls and gender diverse groups. This means working more closely with governments beyond Ministries of Gender and Ministries of Agriculture, with women's and feminist movements, and with human rights organizations.

3. **Bring women's rights out of the dark corner it has been relegated to in the sector.** Religion, patriarchy and colonialism among other social and historic trajectories have shaped women's exercise of their rights. From basic elements such as access to health and education, bodily autonomy, reproductive health, protection from gender-based violence, to more specific areas such as control of agricultural assets, mobility and self-determination, women's rights are openly or silently shaped by social institutions. To date, agricultural development has failed to take a stand on addressing women's rights even in topics that directly influence the outcome of AR4D interventions. It is simply no longer good enough to claim that women's rights exist outside of agricultural development.

4. **Reengineering patriarchal organizations and systems to address gender-based discrimination.** Dismantling patriarchy within organizations requires a clear understanding of structural oppression. The path forward requires a sys-

tematic, sustained and long-term effort, challenging gender norms and stereotypes that perpetuate power imbalances, promoting more diverse and inclusive leadership and fostering horizonal and inclusive relations between different scientific traditions and a wide set of stakeholders. Hiding behind the façade of impartiality and institutional effectiveness arguments in resistance to change only helps to reveal how entrenched and antiquated agricultural development organizations have become.

If we did not believe in change, we would not have put nearly two years of our time into this book. It is time to embrace discomfort and be bold, and we hope readers will join us in this journey to pull women's rights into the center of agricultural development. Thank you for joining us.

Correction to: Gender, Power and Politics in Agriculture

Jemimah Njuki (iD)**, Hale Ann Tufan** (iD)**, Vivian Polar** (iD)**, Hugo Campos** (iD)**, and Monifa Morgan-Bell**

Correction to:
J. Njuki et al. (eds.), *Gender, Power and Politics*
in Agriculture, **https://doi.org/10.1007/978-3-031-60986-2**

The original version of the book was inadvertently published without foreword and incorrect affiliations. The front matter has now been corrected and approved by the author.

The updated version of this book can be found at
https://doi.org/10.1007/978-3-031-60986-2